THE
ADVICE
AGE

THE
ADVICE
AGE

A Letter from the Head of a Financial Services Firm, Circa 2028

ROBBIE CANNON

ON TRACK.
PUBLISHING
Charlotte, North Carolina

ON TRACK.
PUBLISHING

OnTrack Publishing
3540 Toringdon Way, Suite 200
Charlotte, NC 28277
704-562-9163 • info@ontrajectory.com

Disclosure: The information presented in this book is not a substitute for professional advice.

Printed in the United States of America, on acid-free paper.

Library of Congress Control Number: 2023907150

ISBNs: 979-8-9867206-0-9 (paperback)

979-8-9867206-2-3 (hardcover)

979-8-9867206-1-6 (ebook)

Cover and Interior Design: Patricia Bacall • www.bacallcreative.com

"That will mean that finance can be built on all that people do to create value in the network age."

—Jaron Lainer, *Who Owns the Future?*

Here is one possible pathway...

Dedication

To Sheri, my Irmo High School crush, best friend, incredible mother of three, and the love of my life...

FIRM STATS

- $3 Billion AUM
- 15 Advisors
- $30M Revenue
- Avg Fee 1%
- $5M Average Client AUM
- 500 Clients

A LETTER FROM THE CEO

To our team members and clients,

It's hard to believe that a year has passed since I last sat down to pen a "state of the firm" letter. But time speeds along, and while it seems as though January 1 was a week or month ago, almost a full year has gone by. And what a year it has been. A year of astonishing, lightning-fast change for our firm; for the way we do business, for our clients, and for the whole industry.

But before I can set out exactly what that change has been, I'd like first to reflect on the lessons from two of the world's great philosophers and presenters: Copernicus and Jerry Seinfeld.

We throw around the word "genius" with abandon these days, but Copernicus was the real thing. Born in the late fifteenth century, he spoke several languages, understood, and taught subjects as diverse as astronomy, mathematics, and classical literature, and was known far and wide for his brilliance. Not long before he passed away, he published the treatise for which he is best remembered today, positing that the Earth revolved around the Sun and not the other way around. This may seem like no big deal today, but in 1543, when Copernicus first published his theory, it was nothing short of revolutionary.

Jerry Seinfeld, as we all know, is a 1900s-era TV star who created one of the most successful sitcoms of all time—a show that was seemingly all about "nothing." But that "nothing" won record-breaking audiences for nine years before going into syndication and ongoing cult-classic reruns. In fact, it's still in reruns today, decades after it first aired.

So why do I bring these two fine gentlemen to your attention when I'm talking about financial services?

WHAT'S AT THE CENTER OF YOUR UNIVERSE?

The first reason is that this year caps what I would describe as nothing short of a Copernican revolution in financial services. For the first century or so of our current system's existence, the firm was at the center of the financial services universe. The good of the firm and its profitability far outweighed the good of any individual client. This led to all sorts of excessive behaviors, which regulation, the courts, and competition eroded over the course of time. As an industry, we aren't perfect, and there are, and always will be, bad apples. However, blockchain and general compliance technology are providing more oversight so we can act in the best interests of our clients, which is what the Securities and Exchange commission demands of us now. And that is helping us command a decent measure of respect in society again. Just as Copernicus and his intellectual successor, Galileo, upended the world of science and religion during the Renaissance, technology is upending the world of finance today. Clients, not firms and not advisors, are moving to the central spot, and all activity will revolve around them.

You may think your clients are at the center of your universe today but are they really? Think about your revenues—no matter whether you're independent or captive to a firm, or anywhere in between—revenues primarily flow from investment management, not from interacting with your client. You'll get fees whether you talk to your client this quarter or not, and the client will assume you've thoroughly reviewed their portfolio. Maybe you have!

Truth? Until now, perhaps until this very year, you could say that the center of our financial services universe was still our firm based on its ability to manage assets. Decisions as to how the firm was organized and governed, how it operated, how our advisors were compensated, and what information was disclosed worked together to put the firm's interests first. And now, with the changes in technology, access to information, compensation, the use of blockchain, and numerous other developments which we will discuss in this letter, our firm is no longer at the center of the universe. The client now occupies that special position, and rightly so.

We aren't here to serve only ourselves, to make a nice living, and not worry at all about making our clients' money. In just three years, nearly all the Baby Boomers, the largest generation, will have retired, and we've been working to give them a vision of their future as they do. Most of their "book" is now geared toward generating income, which they're drawing from, and any remaining equity will be reserved for the next generation. As these Boomers' books decrease in value, we must lead them to what's next for them and what we're trying to achieve for them.

As you know, we have been in a growth business for a while simply because the Baby Boomers accumulated a historic level of wealth for their retirement. So, what do we do now? Help transition this same wealth in creating their legacy. Good thing we're in the business of helping people. We can no longer say, as Woody Allen famously did, that "a stockbroker is someone who invests other people's money until it is all gone." We now really do work for the people who pay us—our clients, generation after generation.

As it should be and, quite frankly, as it should always have been.

You can call the change a Copernican revolution in financial services, and I think it would be accurate. Copernicus' ideas weren't universally accepted when first promulgated. Far from it. Had he lived longer than the seventy years allotted to him by his Creator, he would have seen the Catholic Church, which dominated the world he knew, rise up in arms against placing the sun rather than the Earth at the center of the universe. Copernicus' ideas threatened everything from the scientific teachings and credibility of the Church to even the way the calendar was established. Incredibly, nearly 100 years after Copernicus first published his theories, the Italian mathematician Galileo, whose astronomical writings supported Copernicus' theory, was condemned by the Roman Inquisition and placed under house arrest.

Two Views of the Solar System: Then and Now

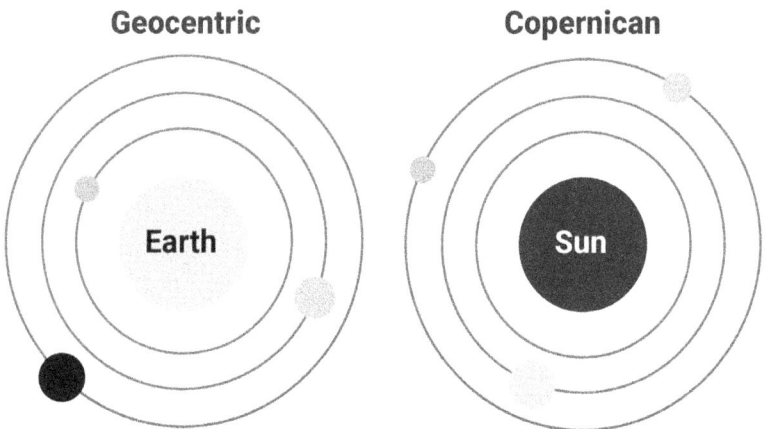

Geocentric   **Copernican**

Earth   Sun

The revolution in thought spawned by these and other Renaissance scientists created a raft of enemies because the power, money, and even glory of those who believed differently were threatened and, in many ways, disrupted forever. We are seeing a similar kind of resistance today. Over the past five years, the Copernican revolution in financial services—shifting the focus from advisory firms like ours to our clients, the individual investor—has generated the same kind of intransigence that Copernicus' ideas faced from the powers that were in the sixteenth and seventeenth centuries.

To put it simply, most firms resisted for as long as they could the idea that they had to recognize the centrality of clients. The industry believed that the major lesson the Great Recession and the Covid years taught was proximity: to *keep the client close meant keeping their respective revenue.* Larger firms realized this axiom and began vertical integration in earnest, one strategy being the "one-stop shop." The thought was, "if we offer every-thing the client needs, they won't leave us." So everyone built the same everything all at once: a modern, one-stop, vertically integrated personal financial services firm. As everyone did, we foolishly followed, assuming if the client could build the same industry to service them, they would have built what we did. Either way, by this new configuration, an interesting new benefit accrued to the financial business: we could all become a "data" company. Why? Until now, firms including our own owned information—data—about cli-ents more from necessity, i.e., regulations, than strategy. Just think about the massive amount of client data (and redundant data) the average financial service

...so everyone built the same everything all at once...

5

company accrues each year. Once we had vertically integrated firms, we wanted more: we went after outside assets. How important was aggregation? The axiom changed a bit. To own the client data is to control the actions of the client. And data monetization strategies went way beyond what asset clients own and the investment strategies to meet them, with perhaps a nudge here or there to achieve sales outcomes. You should be happy to know we never went this far.

Who Is Served?

Traditional   Future

Institution → Transaction → Advisor ⋯▶ Client

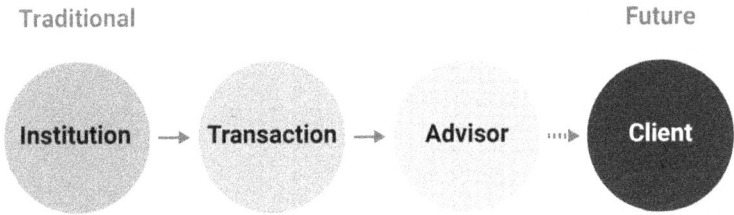

A problem that even we dealt with was what data intelligence to give back to the client, even if the client's assets left the firm. Like a doctor's file, should we share everything, even our thoughts? But our firm chose to be different. Because if the firm or the advisor owns the customer data, then the firm will forever be at the center. How do we know that the client is not at the center? Answer the question: How easy is it for a client to leave advisory firms, comparison shop, and maintain general investment strategies, etc.? The process was possible, but much like transferring medical files, it was possible but very painful. Perhaps we should say assets were semi-portable.

Now, though, clients have control of their data, strategies, planning, advice, etc., which can simply be ported over to the next advisor if the client decides to leave the firm. Data has now switched hands. Since the game had been played on the terms, and turf, of the firms for so long, it seemed downright heretical to surrender that power. And major institutions did what they always have done, since long before the time of our friend Copernicus—they fought back. They made it extremely hard to compare planning offerings, advisors, firms, solutions, etc. The standard line was "you are comparing apples to oranges."

Traditional firms fought back long and hard but, ultimately, fruitlessly. Today, as you know the client runs the show in ways that were unimaginable even a few short years ago. We'll talk about exactly what that means, and how technologies like blockchain and tokenization have sped that process as we go further into this letter. But I wanted to get the ideas of Copernicus into the discussion at the outset because the changes reverberating through our industry today are just as shocking, powerful, and irretrievable as was Copernicus' idea that—think of it—the Earth actually revolves around the sun.

So, what does Copernicus have to do with *Seinfeld*?

Nothing. Which, ironically, is what Jerry has always said his show was about.

If you're like me, if you're flipping through channels in the evening and there's no good game on or any particular movie that catches your eye, you might end up dialing into an old *Seinfeld* episode. It's amazing how well they stand up, even all these years later. The antics of Jerry, George, Kramer, and Elaine are just as fresh and realistic—okay, maybe never exactly realistic, but certainly fun—as they were when the show first debuted in 1989.

But if you take a deeper look at Seinfeld episodes, you'll realize that their world bears shockingly little similarity to our own. How does Jerry communicate? With a cordless home phone that looks more like a World War II walkie-talkie than a modern cell phone. There were hardly any cell phones around when Jerry and Larry David created the show, and the devices rarely make any appearances in the show until later episodes. Can you imagine communicating without your cell phone? Do you realize that in the entirety of the show, not one of the four main characters ever texted anyone? Did they ever search the web? Buy something online? Just saying …

They often got lost because they didn't have GPS. They had to wait in long lines to get tables at restaurants because they couldn't confirm and hold reservations with wireless devices. As Jerry said in an episode when trying to rent a car when there were no cars available, despite his having reserved one: "You know how to take the reservation. You just don't know how to hold the reservation. And that's really the most important part of the reservation—the holding." If they wanted to get together with their friends, they had to get dressed and go down to the coffee shop at the corner because there was no such thing as a group chat. How many face-to-face interactions would Jerry have had with the other three had the show been scripted in the era of modern cell phone technology? Undoubtedly far less. Kramer would never have barged into Jerry's apartment—he would have just shot Jerry a text or made a video call. The characters would have all been isolated in their apartments instead of hanging out together and creating constant comic mayhem.

I bring this up because our lives have changed radically since the *Seinfeld* era. But it's not as if we suddenly went from zero technology to tech 24/7. You know that the transformation

...our lives have changed radically...

was more subtle than that. Things snuck up on us, one app or online service at a time. The same thing is true with our lives as financial advisors. Everything from the way we communicate with our clients to the frequency with which we see them face to face, the ways we receive, analyze, and share market data, and the information we collect as well as the methods for storing it, have completely transformed. This is especially true over the last five years since the Covid-19 pandemic, and, in a more heightened sense, over the past twelve months. Ultimately, what we do for a living is the same: We guide the crucial financial decisions of our clients. But the way we do so today has as little to do with the way we guided financial decisions when Seinfeld and his crew first went on the air. Everything has changed, but often in incremental ways, at least until very recently—if you blink, you might have missed it.

So I'm writing to you today to make sure that, in actuality, you miss nothing.

Put these two factors together—the Copernican revolution and what I'll call the "post-Seinfeld revolution," in which our lives have been technologically transformed in every conceivable way, and that offers a good summary of the state of play of financial services in today's world. Thanks to the interplay of new forms of technology, planning, changes in regulation, and other assorted developments I'll discuss in these pages, there's a lot of great news for advisors. This can all be captured in the phrase I used to title this letter: *The Advice Age*. It's finally here. Our craft is at last to the forefront. Small firms can compete with large ones. Firms like ours compete based on one factor: our advice. We have entered the Age of Advice, and that's what I'm here to share.

As we've seen over the past five years, and as I'll explain in detail in this letter, technology—specifically but not only blockchain—has freed advisors from some of the most onerous, thankless, joyless, and boring aspects of their workdays. I'm referring to gathering information from clients and satisfying the myriad requirements of compliance. We have removed friction from so many points in the process of serving our clients and we have eliminated time-wasting, soul-sucking data-collection requirements through advances in technology that would have been unthinkable just five or ten years ago. As a result, advisors get to do what they entered the industry to do—give advice. If "noise" is getting clients to fill out forms or running every tweet past the compliance department of one's broker-dealer, then "signal" equals time spent giving advice.

Advising is the most satisfying aspect of what we do, it's part of our job title; we are financial *advisors.* We are no longer collectors of financial data, human filing cabinets, or any of the menial roles that traditionally took up most of our days. Instead, we get to do the thing we love the most, most of the time: advise. That means we get to guide, direct, and consult with our clients. It's satisfying, lucrative, creates a sense of usefulness and, best of all, it can't be duplicated by robos who have learned the sorry truth (at least it's sorry for them). *You simply cannot provide advice, simple DIY investment solutions yes, but not advice and not at scale.*

How we got here

Remember, just a couple of decades ago, in the 2000s, our industry was convinced that robos were going to eat our lunch, destroy our businesses, wipe out our fee structure, and end the world as we knew it. Wealthfront and Betterment who? Instead, I'm here to share the good news that advisors, thanks to the advances in technology and other areas that I'll describe in this letter, beat robos at their own game. People don't need a single lever to pull, as if investing were working a one-armed bandit on a casino floor. Instead, they need the very thing we are best situated to provide; advice. We now get to spend most of our time providing advice, and the investing public has discovered a deep need for that precise offer.

Five years ago, we were looking at not only a democratization of investing, which is something we all wanted to see, but also the *clear transition* of our industry. We also missed the reason, at the time. Otherwise, intelligent people were acting as if it were the late 1920s, or maybe the late 1990s, all over again and throwing money at "investments." I have a hard time dignifying some of these plays with that term, as if they were in a casino. And for every investor—I mean, *gambler*—who got famous for making $21 million in a month on a cryptocurrency named for someone's pet, an NFT picture of an ape, or some such outrageous triumph, many thousands of them lost all or part of their investing capital.

But the main issue wasn't people just looking for "get rich" schemes, "dumb" no-brainers or lemming marches. They lost trust. We simply missed it, and

> *We simply missed it, and so did the industry.*

so did the industry. Individuals, our clients, everyone's clients pushed back on the traditional advice industry and even on the Federal Reserve. Of course, we know this now. They were worried that the Federal Government was just printing money and potentially ruining the money supply. Japan was already at

2x debt to GDP and the US had just crossed 1x its own GDP. The Federal Reserve's balance sheet the largest it had ever been. But the biggest issue back then was that investors weren't getting information. Simply, clients wanted specific information on the stability of the US currency if the Government was wrong, regardless of whether cryptocurrencies were the correct alternative. The financial industry's response to any alternatives was silence. With such a crucial question of currency and economic stability, why was it silence? We, at this firm, should have taken a lead role in answering this point. In fact, the popular response was that they couldn't advise on the topic at all. And right here was the moment the industry changed forever. On one of the most critical questions facing the financial services, the solvency of the US government, the industry was silent. The industry clients trusted was silent.

Trust was abused. Enter trust but verify. As a result, a radical shift occurred. Clients were left to figure things out for themselves instead of utilizing the traditional advice channel—the transition began—and some of them even ended up getting their crypto *and* stock advice from conversations in a Subreddit thread or at a party. Or they invested in things that they read about on unregulated websites. They were throwing money at Bitcoin, a defunct video game store, NFTs. They were buying crazy things on margin. It was bound to end badly.

People began making their own decisions, even creating markets that they could potentially benefit from. Basically, they started to opt-out from, and steer clear of, professional advisors. All this activity spawned a shadow industry known as DeFi or decentralized finance based on the new tracks left by the blockchain and cryptocurrency distributed ledgers. This crowd was different from so-called do-it-yourself investors. It showed a new approach and alerted us that the industry was in transition. The industry should have reacted sooner. What started out as a movement turned into an actual industry in the last five years. So, there are now two distinct communities: the traditionalists that continue to use and trust advisors, and the opt-out, or DeFi, group that turned away from advisors when they didn't adapt to new realities.

Many of these individuals soon concluded that "robo-advisors," digital investing services based on algorithms, which offered, at least in theory, the ability to invest quickly without the "dreaded" relationship, pointed to dead ends. Indeed, they were simplified investment solutions, but they were just a map and an arrow that said "go that way." I can't even remember if, when we set ours up, we told our clients when and how to use them. But like ATMs, automation came at a price and obscured what real advice looked like. We knew this aging robo technology was actually just slick new marketing of an old investment idea. Robo tech looked like a spaceship from the future, but the industry knew it was just a beat-up station wagon being marketed as a spaceship. It made small accounts easy to administer. It was somewhere to put the nephew's five-thousand-dollar IRA. Wow, looking back, we should have seen the mixed messaging and confusion we were issuing. We should have seen the transition coming. Ironically, for all that robos and crypto spaces were not, they were the necessary building blocks that led to where we are now.

Let's go a little more deeply into the various areas that have either facilitated or benefitted from the changes of the last twelve months and, more broadly, the last five years. And we'll start with one of the biggest shifts: how blockchain technology changed everything.

For the Advice Age, advisor technology must reconfigure.

Technology Configuration — What's Coming

	Traditional	Imminent
Backbone	Daisy Chained	Blockchain Wealth Passporting
Ownership	Firms Control Relationship	Client Controls Relationship
Client Interaction	Semi-portable	Portable
Financial Data Access	All-Access	Permissioned Access: Vaulting, Services Planning, etc.
Financial Business	Advice Plus	Advice
Fee	AUM	Ongoing/On-demand

HOW BLOCKCHAIN CHANGED EVERYTHING

I t's hard to think how our firm, or any firm, could operate on a day-to-day basis without blockchain technology.

It is a fundamental change that is suddenly everywhere, like the smartphone. You might not notice the change has happened until it's impossible to imagine functioning without it. Pulling out a paper map for directions or seeking a pay phone to call home when you're delayed seems completely ridiculous to us today. These kinds of behaviors are confined to old movies now. But just fifteen years ago, they were part of our everyday lives.

Blockchain has come into our lives and the life of our firm with incredible swiftness. Five years ago, most firms couldn't imagine a place for blockchain outside of the Wild West of the crypto world. The technology was too closely associated with the kind of speculative, unregulated trading, advice, and feuds around cryptocurrency as chronicled on op-ed pages and by influencers on TikTok and Instagram.

Thankfully, blockchain's use in the medical field to manage electronic health records inspired a similar innovation in financial services. Using blockchain to store, manage, and permission the entire financial history and preferences of individual clients left the field clear for a redefinition of what services firms can and should offer. It also turned the relationship between advisor and client on its head, for the better.

In this section, I'll reflect on how blockchain provided an opportunity to reshape the role of the advisor, the client, the firm, and its competition in a very short time span.

An obvious and crucial starting point is the amount of paperwork blockchain removes from the equation. Digitization is a part of the technological revolution that firms both large and small have contended with as the internet takes a more central role in our lives. Financial services had long been a holdout, stubbornly clinging to the idea that the kinds of sensitive records and identity verification required for our transactions needed to be on paper and in person.

Our youngest clients may not remember the stacks of paperwork, forms to be reviewed, signed, and filed, and time spent waiting for the even most basic of transactions, opening an account, to complete.

But the rest of us sure do. For example, just five years ago, if you were a high-net-worth individual who did private real estate deals or licensing through a network, you had to constantly fill out accreditation paperwork to qualify and meet certain income guidelines and SEC criteria. Processing that paperwork could take days. That meant you had to get accredited repeatedly, every time with every investment opportunity. There was no stamp that said, "This person is accredited. " What a waste of time!

But now, once you put your financial history on blockchain, you're instantly listed as an accredited investor or qualified purchaser. On chain, that single element of financial information facilitates a private passport that denotes a larger, verified data store, which gives you a lot less friction against doing deals because it establishes your net-worth credibility right away.

While the traditional, old-school, station-wagon way of validating eligibility was a credit check, a certification, etc., blockchain creates new efficiencies. Now, thanks to blockchain, you're represented by one number, and you control and permission that data and information.

Wealth Passporting: How Client Information is Stored and Shared

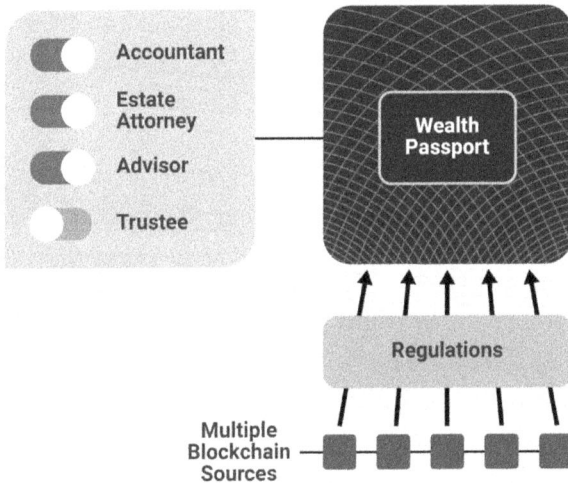

In one fell swoop, blockchain didn't just reduce the amount of paperwork involved in our client interactions, it eliminated it. Every interaction with the client is not only completely verified it is also recorded in the blockchain's immutable record. All those friction points, opportunities for human mistakes, and waiting periods while documents are reviewed are gone. Better for the client, better for us.

A jumbled bunch of papers became a complete, unchangeable, and unquestionable digital record. And this digital record?

It is not sitting in a file cabinet somewhere, or on an encrypted server, instead, it remains securely in the client's hands. (See developments in China for the opposite outcome. There, every citizen is assigned a QR code which the government can track and even tie to facial recognition software.)

The realization that clients, not advisors, now have primacy of position caused collective shudders across the world of traditional financial advisors. However, many advisors and firms recognized the enormous benefits of this swap and benefited greatly.

Giving clients a central role in managing their financial history and data is not the same as relinquishing control of that information. Instead, it gives both sides of the advisor-client relationship more freedom and flexibility to customize what the relationship can deliver.

Blockchain has given advisors that time back and then some.

All that time wasted on data collection, verification, and control has been returned to us. Advisors can use it on much more valuable activities.

Part of what sets financial services apart from most other fields is the amount of specialization and depth of knowledge advisors have at their fingertips, and the near impossibility of navigating the field without guidance. Clients come to us for that guidance and education and, as I've laid out elsewhere, that face-to-face time is irreplaceable.

Blockchain has given advisors that time back and then some. Freed from the drudgery of shuffling papers and chasing information, advisors returned to their eponymous function, i.e., giving personalized advice and working closely with clients.

As you know, in our firm, our advisors now access the financial data new clients have input and permissioned on the blockchain and can have productive meetings to discuss future steps right off the bat. Advisors don't have to waste several meetings reviewing financial history and hunting down missing documents, and clients can feel that their time with us is well spent from the start. It isn't just the advisor's time that is saved, clients no longer have to wait days or weeks for a notice of an approval. Instead, within seconds, it becomes clear what clients can do with their money based on the records they've shared.

It's also important to recognize that client permissioning is not irreversible. Clients' control of their information and data stored through blockchain is the final step in personalization. They can choose to de-permission their data. They might want to change firms or make their financial management more hands-on where advisors aren't automatically making all the decisions for them.

More than ever, this requires all advisors to be at the top of their game. Not only have they been restored to their rightful roles as financial experts, educators, and advisors, but they also have to wield this responsibility well or risk losing their clients to a firm with more responsive and able advisors. Blockchain's transferability and one-size-fits-all qualities mean clients are no longer held hostage to a firm just because they have spent a long time there. The data controlled by firms that once held many clients back from considering better fits for their needs is no longer an encumbrance, as clients hold complete control over their financial information. As Jaron Lanier argues, it's important to realize who owns the future.

In-depth face time, planning around financial goals, and maintaining a personalized relationship with each client were luxuries most firms could not afford as they drowned in paperwork and chased down scattered financial records. Now, these elements are advisors,' key priorities as they always should have been.

Speed, Stability, and Power

The longevity and immutability of blockchain is another key asset in the advisor-client relationship. When advisors perform well, bolstering the firm's reputation, clients' children, grandchildren, and great-grandchildren are motivated to remain with the firm. Their financial records are seamlessly incorporated through blockchain technology. This has helped smaller firms maintain and even grow their client base. As older clients retire, pass away, and become less actively engaged in their day-to-day financial management, they are being joined by younger generations who have been raised in a highly-digitized environment.

Younger generations have radically different expectations across the board than their parents and grandparents. Their financial goals, the role of financial advisors, and even the way they interact with firms managing their assets have all been shaped by exposure to the internet and a lack of exposure to financial education and resources. As a result, younger clients desire easily accessible and intelligible, not to mention immediate, overviews of their finances. There were fears that this generational shift would cut the advisor out entirely but, as we have learned, it has simply helped advisors showcase their strengths as educators and financial counselors.

With blockchain cutting out the fuss and paperwork that daunted and alienated young clients in the past, they have moved directly to building a relationship with their advisors and learning how to navigate the financial world with their advisors' guidance. As advisors report back to us and other respective firms, we are reshaping our priorities and services based on the changing financial habits of each generation. And best of all, because blockchain creates a permanent, immutable record of every client transaction made, it remains a constant point of reference in our analysis and planning.

Blockchain also smooths sudden financial transitions. In the event of a tragedy, which would typically throw finances and family life into disarray, surviving spouses and children do not have to spend grieving time doing financial detective work because, again, all this information has been consolidated and permissioned to whatever degree the client wanted through blockchain. The acquisition of assets and debts in a new marriage,

> *They created an environment that sold your data...*

or their division after a divorce, are also seamlessly managed and recorded through blockchain.

The systems of the past, like virtual vault services and aggregation technologies, didn't make individuals feel protected in times of need because they only reinforced aged client data models. Although these entities housed your data and helped you organize your financial picture, their actual goal was to display it to interested parties and monetize you. They created an environment that sold your data to anyone who wanted access, trampling over your privacy. While the old models started with data access, the new ones start with a shift in data ownership, putting the client in control. Blockchain allows clients to do the business of organizing their financial affairs when family tragedies occur without having to worry about uninvited people showing up to the funeral. See Kinnect.

The speed with which firms have had to incorporate new and existing technologies into their client services is breathtaking, and there's nothing to indicate things will slow down any time soon. The software programs that firms create or use, their program management systems, and their entire digital presence are being constantly re-tailored to meet clients', advisors' and competing firms' demands. As you know, our technology stack still consists of custody and client relationship management systems (CRM), and advisory product solution platforms. Of course, planning platforms have been transformed; but more on that later. Fortunately, blockchain has made accessing and tracking client data seamless and almost instantaneous, so advisors can spend more time educating and analyzing the complex optimal interactions of the market, taxes, estate, banking, and insurance, and then use what they learn to make better decisions with their clients. Firms now can focus their energy on providing solutions and platforms that are accessible and user-friendly for both advisors and clients.

The post-information gathering age, the Advice Age. The pre-Copernican world valued sales over advice, the barbarian over the philosopher, growth over quality. We measured books of business in AUM; now we look at and evaluate advice-quality scores.

The consolidation and simplicity that blockchain has granted financial management firms have assuaged one of the fears that haunted the field not long ago—the rise of robos and the decline of traditional models of financial advising (that is, the need for humans) in the model. We know now that this fear was unwarranted. Not only have robo-investing platforms taken up much a smaller share of clients than anticipated, but also leaps in technology and AI and blockchain in particular have allowed advisors to bring many of the supposedly alluring aspects of robo-investing into a blended, digital-human advice model. In fact, robo-advisors should have been called robo-investors. At this firm, empowered with machine learning, our advisors' scores are at industry highs. Prediction over description, precision over "off-the-shelf," trusted but verified, and advice over anything less.

Because in the past, the false narrative and erroneous assumption was that advisors are basically sales representatives, relationship monkeys, and even well-known institutions like Vanguard, Morningstar, and Envestnet seemed to suggest this at times. And if you were not tethered to an advisor, you could make choices that directly and objectively benefitted your financial future, and it really wasn't that hard. Furthermore, you could make these self-advised decisions from the comfort of your smartphone or laptop at home, all for low or no cost, making the barrier to entry nonexistent. Truth to tell, these assumptions are correct, if and only if asset markets are rising smoothly. It's easy to feel smart in any bull market!

As we know, these assumptions were all part of empowering an opt-out that the industry didn't understand. And the bull market, of course, did not last forever, which consumers didn't understand until they lost lots of money. Successful financial management can be just as delicate a balancing act as performing heart surgery. No one in their right mind would set foot in an operating room and claim to be qualified just because they are a hardcore fan of the TV medical dramas *ER* or *Grey's Anatomy*. So why would people who get their financial advice from blogs or Subreddits suddenly take it upon themselves to preside over their entire financial future (e.g., goal management, product selection, market dynamics, investments, etc.) and that of their family through a robo-investor firm? In fact, our largest influx of new advisory clients is coming from the "DIY" space.

Opt-out came about as a reaction to the way control was set up by the advice industry. Advisors didn't specifically have an advice role—they had a financial oversight/control /discretion role and moved toward the AUM (assets under management) model to give clients ongoing advice—and in the process, smooth out their own income stream. The AUM model worked well: The industry determined how fees got paid, how data was handled, and the way financial advice worked. But in the neo financial services, DeFi and blockchain space, there are coaches who offer answers for an hourly rate and put their expertise up for bid—and *never* control a lifetime of financial data. It took traditional advisors years to figure out that the DeFi space was working to undermine them by giving both discrete or "on-demand" and continuous advice. The industry has had to catch up. Now, the discrete and continuous advice model is where the industry has been forced to reside, away from its traditional AUM only model.

Blockchain technology, by consolidating all personal financial information and history, down to each individual transaction, creates a digital, encrypted "file" of each client that can be interpreted and analyzed in detail. This means that advisors can empower personalized decisions and solutions instantaneously, with no waiting period. This data is stored on chain by the client, meaning that the degree to which advisors are involved in day-to-day asset management is decided by the client, not the

The industry has had to catch up.

firm. They can de-permission and permission data whenever they want. Remember the transaction controls around credit cards? Finally, this means that face-to-face time with advisors is time well spent planning, setting, and reaching goals, and personalizing management instead of wasting time and money for the first few sessions gathering paperwork and information. The new repercussions of a new science, a new center, a new age.

THESE DAYS, WE ACTUALLY GET TO ADVISE PEOPLE

In any other profession but ours, the work of an advisor would seem pretty self-explanatory. Advisors advise. That's what they do all day.

Think about the right-hand people who advise presidents, prime ministers, or monarchs. Whether fictional like Merlin, biblical like Joseph or Nathan, ancient like Aristotle, or modern-day like "Bush's Brain" Karl Rove, advisors give counsel, review concerns, help set goals, and provide expertly informed analysis. Presidents, prime ministers, and monarchs listen to this sage advice and rely on it. Without good advice, they wouldn't be able to function. In fact, their entire tenure as leaders would crumble without able advisors. The power is in the hands of the leader. Advisors advise; leaders make the decisions. And somehow the world continues to turn.

Then you come to our world, the topsy-turvy world of financial advisors. We advise, of course, but it seems as though we spend most of our time doing anything *but* advising. As we previously discussed we're collecting data on forms, wrestling with compliance, coping with office politics and talking nervous clients off the ledge in fast markets. We're also thinking a lot about succession and who is going to take over the practice, or whether we need to sell it completely to the latest private equity roll-up. The replacement rate for advisors has been low. In fact, there are fewer advisors today than there were just five years ago. So, it doesn't look like we spend a ton of time actually advising, do we? Or at least that's the way it has always been.

That's why I'm so pleased to say that we've entered what I call "The Advice Age." Due to the breakthroughs created by new kinds of software, and other topics we'll discuss, advisors are enjoying an unprecedented ability to spend their time doing what they should have been allowed to do all along: focus their time and energy on providing advice and serving clients. As I said earlier, advice is now both on-demand and continuous. We're answering a question about what a client should do with their sudden $50,000 bonus, which mortgage makes the most sense and how to hedge it, and, based on market changes and statistics, whether they should rely on the same portfolio for income. All done without the limits of the old model.

> *...we've entered what I call "The Advice Age."*

It sounds so obvious; we're advisors, so we advise people, right? Not so fast, pal. Our clients trust us, but that is a double-edged sword. We have created efficiencies from many of the time-wasting activities, but instead of better and better advice, we have seen some practices turned more or less into lifestyle businesses, going through the motions to drive an income supporting a lifestyle. Dinner, golf, wine event, chat about investments and repeat. Do you know anyone like that? Well, this has begun to change, and in a very powerful, dramatic, and industry-changing way.

Let's take a look at the way it has always been. The first meeting between the advisor and client is a meet-and-greet. Do the advisor and the client pass the smell test? Does he or she give off a sense of trustworthiness? "Jim" at the club said his guy was great—is he? That's what both the advisor and the potential client are trying to decide in Meeting One.

Okay, good news. You both passed. In fact, you're hired. Meeting Two is when the client brings the proverbial shoebox full of records. Brokerage statements, insurance policies, wills, you name it. You feel like a clerk at the Department of Motor Vehicles trying to decide if the guy on the other side of the counter brought in enough documents to qualify for a real ID. And inevitably, just as at the DMV, there's some kind of document missing. So, you get a limited start on your planning process. But to do any meaningful work for the client, who hadn't realized that a given document or statement was of critical importance, you have to wait. Back home, the client goes to investigate so you can have what you need to dig in and get a complete picture of their situation.

A couple of weeks pass, maybe a month if the client is working long hours and can't break away, or if getting you the right documents falls into the category of "I'll do it when I get around to it." You follow up. It's important but not critical in their minds. Hey, you aren't fixing a toothache that demands immediate attention. All you're doing is helping guarantee their financial future. Not that important, right?

Finally, the client shows up for Meeting Three, the correct and necessary documents in hand. Hallelujah. The rest of the meeting is devoted to showing the general planning roadmap and small talk: *Where do you think the market's going? What's going to happen with interest rates? Who do you like in the next election?*

How do you like the Cowboys this year? And people call this "building rapport." Some advisors I know call it a complete waste of time; an hour devoted to the handoff of a document that should have been available weeks ago.

And then comes Meeting Four, in which the advisor can finally ask the client key questions based on the complete picture of their finances, insurances, and the remaining documentation that has taken a month or two to gather: *What are your goals? When do you want to retire? What do you want your retirement to look like? How's your health? And by the way, what's up with your kids? Are they mature or knuckleheads who don't understand the value of a dollar?*

We're really looking at Meeting Five before the advisor can truly do what their job title implies—give advice. All that wasted time, energy, and momentum due to the chase for unending documents and statements. Is that what advisors were meant to do? So you can look across the desk and say, "Mr. Smith, I still need the statement from your last advisor." I don't think so, but that's how things have always been.

Until now. Unlike the marathon I just described, the workflow is as clean as a client sitting with their high-priced NYC attorney, who is prepped and ready to go.

Today, technology means that the endless search for documents is no longer the bane of our existence. Everything about the client is up on the chain—bank accounts, brokerage accounts, insurance policies, health data, contracts, you name it. Okay, you still have to ask whether the young adult kids of the client are solid

> *Everything about the client is up on the chain...*

citizens or knuckleheads. And yes, not every piece of data makes it onto the blockchain, but most do. Which means that if you pass the aforementioned smell test in Meeting One, the client can permission all relevant data for you right then and there. At this firm, we review and prep, so our advisors can get down to advising in Meeting Two instead of waiting like in years past. And it's not just the data that is going to be permissioned. The planning visualization that the data builds can be viewable and permissioned to be editable by the professional. So we can review past planning solutions and diagnoses with the knowledge that someone might review ours.

Ain't life grand?

Let me speak directly to your clients for a bit.

Every decision you make about your finances at this moment will continue to impact your well-being for decades down the line. That is where the crucial role of the advisor comes back into play. Think golf: any slightest change of the club head upon hitting the ball can send your shot significantly off target. The relationship you have with your advisor is not just between the two of you, it's triangular, amongst you, your advisor, and your finances.

It's also a symbiotic relationship. If you and your advisor are communicating clearly about your goals and what you want from having your finances under their purview, it's easier to find your way. If you aren't clear about what you need, or your advisor isn't tapped into the complexities of both the market and your situation, you may find yourself in a tight spot.

Today, the advisor and client stand on an equal playing field, which would have been impossible without two central technological mini revolutions in financial services, blockchain and personalized planning innovations.

Bringing in your next generation of clients

With this symbiotic relationship in mind, let's get back to our advisor.

Earlier in this letter, we covered the incredible transformations blockchain brought the whole financial industry. Blockchain not only slimmed financial services down into a sleek, personalized, and high-tech field but also brought advisors to their proper place— at the client's right hand: their financial sage.

The sheer amount of time regifted to the advisor has permitted them to reconnect with individual clients and in doing so, rediscover one of the special gifts of this position. No surprise. The knowledge economy was bound to do it eventually. Being able to meet on an individualized level with clients brings that extra level of depth and trust that only human relationships can convey. It also highlights another aspect of advising which had been previously buried in the piles of paper advisors had to sort through—education. Financial education is paramount to our advisory relationship. You perhaps have seen some financial firms that used this efficiency incorrectly— to drive more leisure, three-day work weeks, sub-10 handicaps, etc.

Education may not be the first thing that comes to mind in the financial services world. In fact, financial literacy is a constant aspiration. As clients become younger and other firms and we take them on in a greater volume, we have recognized an enormous knowledge gap in this new client demographic. It has become clear that the complexities the services firms offer, and the intricacies of the market and financial planning, are impossible to interpret alone. And that's where an advisor steps in. Financial literacy has taken a leap forward, and the more financially literate people are, the more advice is actually valued. Like clergy of the past responsible for the religious practice and growth of the entirety of the family, our firm has established, Level, our family financial literacy program, designed to empower our clients: everyone, every demographic. Clients are automatically enrolled and Level elevates even the most simplistic understanding to informed viewpoints.

Financial Education—the Client's Journey

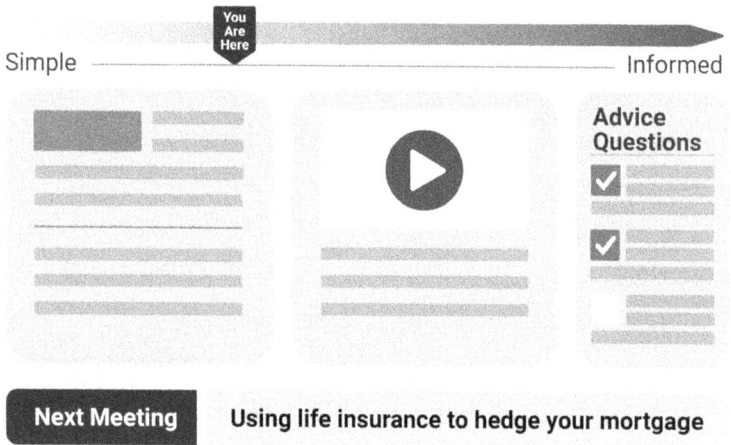

There's a new scenario in which advisors are especially exercising their strengths as financial educators. Younger clients are coming from more diverse backgrounds these days—not every new prospect is a fifth-generation heir named Chip, who lives in Greenwich and summers in Maine. Many of our younger clients are the first in their families to safeguard their financial futures by entrusting them to advisory firms. In fact, according to an old Fidelity study, approximately 88% of American millionaires are self-made.) As novices in the field, they rely on the expertise and knowledge of advisors to guide them through the labyrinth of decisions that awaits them. If it's the "Advice Age" for the advisor, it's also the "Age of Impatience" for the client. Thanks to the internet they are used to instant access to information and nearly instant responses to their messages. Gone is the arduous task of assembling all one's financial data and bringing it in the proverbial shoebox to a new advisor.

As we've seen, blockchain has given our younger clients, as well as the children and grandchildren of long-term clients, a new kind of agency. Loyalty to a financial services firm, ours or any other, has become firmly based on the competency and quality of advisors. Dissatisfied or impatient clients today move on swiftly to a better-fitting firm, just by de-permissioning information on their blockchain and giving the new advisor access to the data. So, the freedom from data collection that always plagued advisor and client alike actually requires us to step up our game. Because it's frictionless to switch firms, something that was never true in the past, clients expect more from us. And rightfully so. If we can't provide the advice they want when they require it, they'll be out of the door. The freedom to spend the bulk of our time advising is also an obligation to spend more of our time advising. And there's nothing wrong with that.

...clients expect more from us.

More time to advise means there are more types of issues on which to advise. Thanks to blockchain (and the planning developments we'll touch upon shortly), every kind of product and financial goal is now on the table. We will talk about planning for retirement, of course—and we can also address more quotidian needs, like saving for end-of-year vacations and paying off student loans. We have the luxury of time to educate our clients as to what a strong, well-balanced portfolio looks like. Previously, these educational conversations were previously unfeasible. So much of our time was devoted to collecting data, begging clients to send in missing documents, and dealing with unending compliance concerns.

Education has become central for advisors in another way, too. We get to spend far more time these days poring over the portfolios of individual clients. We have the time and technology to see patterns, strengths, shortcomings, and optimal changes. We can see things we never could before and advise accordingly. All that information we collected in the past, on physical forms, locked away in actual or digital file cabinets? We were really playing defense so we could fend off unsuitability lawsuits, compliance audits and the like. Now, because all our data is available online, we can make use of it and study our clients, understanding them and their needs as never before.

Bringing in your next generation of advisors

These changes have led to a different set of hiring practices. We no longer look for the traditional junior advisor—the young man or young woman, sales- oriented, hammering the phones, making the proverbial hundred cold calls a day, searching like Ahab for a whale to bring in. We're no longer as dependent on the process of turning a bunch of juniors loose on friends and family, and seeing which ones emerge from the Darwinian process of elimination—hire ten, keep two.

Instead, we're bringing in more and more young specialists. We're hiring insurance analysts. We're hiring investment analysts—even data scientists and ex-educators. What is the thinking driving this revolution in hiring? Data. Information. Practices, historically, were built two ways: 1) with great sales-people, 2) with a room full of investment professionals. Now, it's all about experts in different fields with varying experience and backgrounds.

Innovation and insight happen at the intersection of really smart people with different backgrounds, not only for the firm but also for the client. These disparate parts make up a new advice engine. It's collaborative planning. We are using that new engine to empower the clients we are entrusted with. The "Age of Advice" will be the empowerment of the client. The client will walk out feeling like not only that they are in control of their data but that they now function at a higher level in terms of the advice that was given. Education given by experts equals client empowerment. Trust is great especially when it's partnered with empowerment.

We know so much more about our clients, their needs, their aspirations, their portfolios, their insurance, their banking, credit needs, even their financial acumen. Again, this kind of information used to be locked away in file cabinets, with those documents yellowing and becoming out-of-date. Today, we know exactly what's going on with each of our clients, in every aspect of their financial lives.

The smart play, we decided, was to bring in bright young analysts in all fields alongside our data scientists to examine and make recommendations to our advisors, who can then absorb and pass along that advice to our clients. It's the grandmaster with the AI, the physician with the diagnosis algorithm. It's finally here, it's financial advice. The driving idea is to create growth in the firm by providing greater value, in a highly personalized, data-driven way. The more value we create, and the more value clients perceive, the more likely they are to refer other high-net-worth individuals. We already had measurements for investment performance, assets under management, risk tolerance. We just added advice-quality ratings. More than just Betas and Alphas, passive or active investment strategies, it is about the actual quality of our advice: its ability to help clients achieve goals, navigate financial complexity, wellness. It's a true rating of advice, beyond customer satisfaction, optimal decision-making. In a competitive, market-driven way, the rating is the determination of superlatives. It has taken a while but now it can determine good advice and good advisors.

Retaining clients who used to walk

Here's another consideration that requires attention, even if it isn't politically correct to make the point. With our older clients, and even with couples in their forties and fifties, we often have primary contact with the husbands. I know society has changed and that we're on much more egalitarian footing than ever before. I'm aware of all that, and as a husband and father, I'm glad. And yet, the news that men and women should have equal access to knowledge about, and plans for, investing, insurance, and retirement still hasn't reached most married couples we serve. Today, increasing amounts of global wealth is in women's hands, and that has become a significant challenge for the industry. It's a solvable

I know society has changed...

36

problem that we could, and should, have fixed back then; way back then.

In the past, who mainly came to see us? The men. Who signed off on the advice we offered and the plans we created? The men. With whom did we enjoy golfing events, dinner, drinks, phone calls? The men. Who make the bulk of the decisions about where to put the money, why, for how long, and when to sell? The men. It's not because women were incapable of such decision-making, and it's not that we won't talk to them! But for whatever reason, paternalism, sexism, or just something built into our DNA, the husbands are traditionally the ones who handle relationships with the family financial advisor.

And by the way, the husbands are usually the ones who die first.

Consequently, some widows who come into our office after their husbands have passed are in a quandary. They have no idea what investments they own. They don't know where the accounts are. Sometimes they aren't even sure who their husband's advisors were. You might think this sort of sexist scenario went out of the window in the 1970s. Nope. The husbands might have been motivated by a desire to insulate their wives from the ins and outs of their financial portfolios, or maybe they thought it was boring or that it was "man's work." Maybe they had something to hide. But ultimately, the guys didn't do their better halves any favors by keeping them in the dark.

And now, the widow is going through shock, loss, and grief, and maybe exhaustion following her spouse's lengthy illness, and she has no idea what's flying in her financial life because her husband didn't tell her. In the past, advisors were often put off by her "incompetence" and grief, and in years past, these women left their husband's advisors in droves.

Once again, friends, technology changes everything. The husband might have been best golfing buddies with his advisor; that's hardly unheard of. And he may never have spoken one word about financial issues with his wife, but now he's gone, all that vital, juicy information lives on for the advisor and widow to review together. No longer must the newly-minted widow go on a proverbial paper chase, trying to piece together a picture of her financial life without her husband. No longer must she go through the drawers in her late husband's home office desk, frantically trying to determine what insurance exists, who the broker is, and what steps she needs to take now with the IRS, the state, and so on. Instead, she simply comes into any firm and sits down with any advisor who has instant and complete access to all the information necessary to properly advise her about what there is, what needs to happen, and where she goes from here. Yes, it's just a "vault," but the widow owns it and controls it. She is in control of it all; it's a Copernican revolution. The Age of Financial Advice is for Everyone.

I fear the last few paragraphs describing the sorry state of service to widows read like a *Ladies' Home Journal* article from the 1960s. The reality is otherwise. It is a constant source of shock and chagrin to our advisors when a client dies, and the surviving spouse (usually, but not always, the wife) visits, perhaps for the first time, without the slightest idea of what's where and what's now. It happens every day in every financial services firm in the nation. Or, I should say, that's how it *was*. Thanks to tech advances, the un-merry widow is becoming a phenomenon of the past.

So, what does it mean, the "Advice Age?" How do we summarize the topic and move on to the other developments that have so radically changed our industry in the past year?

In the not-so-recent past, "advice" meant touting a portfolio that the firm was "pushing" on the permissioned list, not unlike a diner owner telling the wait staff to push the "special." It meant throwing darts in the dark because we didn't know everything we needed to know about our clients. It meant we were judged not by how much value we created but by how little money our clients *lost* in any given year. Our reputations weren't based on our advice; they were steeped in basically how much money we had under management, essentially, how much money we made them. And if we played conservative with their investable assets, they probably wouldn't lose much; they might even make a little something over and above the fees we charged, and we wouldn't find ourselves with suitability problems.

Now that we have access—dare I say *unprecedented permissioned access*—to real-time data regarding our clients' financial lives, we can help them really *plan*. The investments, banking, and insurance products we recommend are now a function of the advice we give. We're actually helping people get on track or stay on track with regard to their financial future. We can look at their portfolios and their lives, not in a detached, discrete investment-by-investment manner as in the past. Technology now enables us to "level up" in the way we see, act on, and share the big picture in ways we never could before. Try comparing a car from the 1920s to one today.

> *We're actually helping people get on track or stay on track with regard to their financial future.*

Sure, they're both *cars*, but the new one has degrees of sophistication unimaginable one hundred years ago. Today's car now literally drives and parks itself. In this new environment, we'll be judged on how well we assemble and act on that holistic, 360-degree viewpoint. In short, we can do what our job titles have always implied that we do: *advise.*

Again, ain't life grand?

PERSONALIZATION THROUGH SOFTWARE

The impact technology has had on the financial advising industry is profound. But until recently, tech has fallen short in a few significant ways.

For one, until now, not one of the popular versions of planning software in use has been able to track all a client's goals at the same time. Let me finish. The software can monitor individual goals, like retirement plans or savings targets for the kids' education. But real investors constantly juggle various purposes at once. Not that there was good progress. No company has offered software that can chart the spectrum of goals simultaneously, the way real people aim to do. A quick aside about goals-based planning: This firm believed that McKinsey got it right almost a decade ago when it argued that goals-based planning would be pervasive and dominant by now. A lot of firms were not only using goals-based planning but firms like Horizon, SEI, and Brinker were also offering products designed specifically for the life cycle of a goal, from accumulation to completion. The industry now uses goals-based planning as a table-stakes planning medium.

However, there is an issue of software ownership. The industry heavyweights have traditionally maintained possession of the planning technology advisors use *as well as all the information it collects from clients.* This tight grip over software ownership has allowed the big companies to maintain control over advisors, including their offerings, pay, and mobility. By extension, they exercise that same control over clients.

All that has changed in this era of bold innovations. Last year, we introduced new software that restructures these controls and other fundamental aspects of financial management. Above all, machine learning allows all an investor's progress toward their goals to be tracked in real time. It will be refined a bit later this year and will be available for individual advisors to purchase. It's impossible to overstate how much these software changes have already affected our advisors and investors and our approach to advising. Tracking the full gamut of clients' goals allows our advisors to work closely with investors in ways that have never been possible before.

Tracking Clients' Progress On-Screen

Machine Learning Course Correction

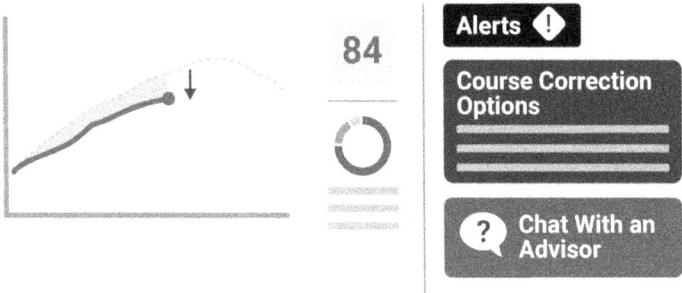

84

Alerts !

Course Correction Options

? Chat With an Advisor

As the new planning software becomes more available, the relationships between the advisors who own it and the firms they work for will also adjust. From what I have been told, this software works a lot like those that owned a seat on any trading exchange. Advisors will have more leverage because this "seat license" allows their planning software to reside outside any of their financial firm affiliations.

This amounts to two Copernican revolutions at once. Before the year is over, these adaptations will bring the same breadth of technological innovation in the world of financial advising that we have seen in other fields over the past two decades.

Some improvements are already well underway. As we have noted, our advisors are using on- the- chain wealth passporting, which is on course to revolutionize the advising workflow. Investors' life circumstances are constantly evolving. And their investment goals are shifting with those changes. To craft an investment plan that responds to those changes, our advisors not only need the data that passport gives them, but they

Advisors will have more leverage...

also need constant updates about the significant new events in an investor's life.

Our new software has rounded out the sweep of innovations. So far, our advisors are thrilled with how the new software is working.

To illustrate just how it has improved their work, let's take a close look at the process an advisor with one of the mega-financial services companies currently goes through using the best software currently available. The advisor manages about two hundred clients, with assets totaling $200 million. To simplify things, let's say that each of those clients has five goals. Retirement would be a common one, then maybe buying a second home, plus other goals down the list. In total, the advisor must track roughly one thousand goals for all their clients.

Now remember, different generations and demographic groups have unique aspirations. So, the advisor must keep stock of goals that differ a great deal. They must understand where each plan starts, where it is finishing and its status at any particular moment. They would also need alerts about when a goal goes off track and insights on how to get it back on. Even the sharpest of advisors could not juggle that volume of information and the accompanying web of figures, especially in fast markets. Good software should be able to help.

The software most advisors regularly use in tracking goals lacks the sophistication to do that level of calculation. However, their software does perform some valuable functions: If the investor crunches in the correct numbers, the software gets the basic stuff right, like an investor's net worth or spending power for a given year. If a client wants to make a particular type of investment, the technology can help pinpoint some good options. But none of the earlier versions of the software has been capable of deciphering much of the important stuff. It can't specify the tradeoffs a client will have to make to reach their goals. Further, it can neither tell what getting off track really means and nor what to do with excess when you have it. It knows what you have and when you have it, but it is silent on crucial decision points like shortage or surplus. It simply plays the large number game with the markets. We can do better and should do better, so we do.

Goal "excess" is an interesting concept. For example, if our client has a higher probability of success to reach one of their goals thanks to a positive sequence of returns, the software can highlight the difference between the higher value and the "projected" value and suggest more efficient uses of resources. That is, the client can take some of the money they would otherwise allocate to Goal A and use it to achieve Goals B and C instead. Naturally, our advisors are there to deliver these kinds of assessments. But getting access to new data about the investors' goals can be a big assist in the advising process.

Excess As Strategic Advantage

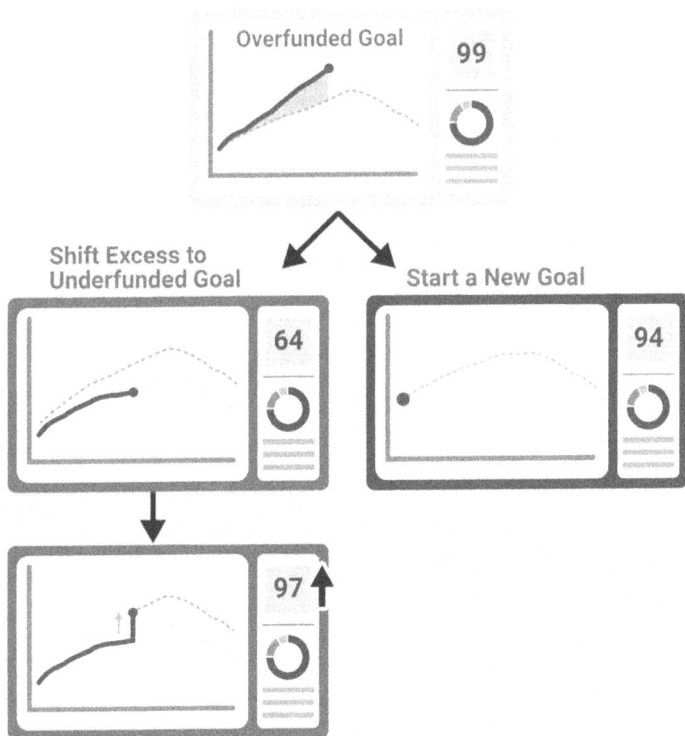

Overfunded Goal 99

Shift Excess to Underfunded Goal 64

Start a New Goal 94

97

The new software's ability to record the range of a client's excesses and shortfalls in funding specific goals is profound. It also has the capacity to capture changes in those goals as they occur in real time. Let's look at the case of some investors we're working with, a couple in their thirties who are investing together. They want to save money for college for their school-age kids. One of them wants to launch a small business. They are also expecting a hefty inheritance from one set of parents and want to invest it. Various critical points in each of these goals require action, and the new software tracks all the waypoints. As new plans come along, for example, financial adaptations brought on by the birth of another child or a job change, the software registers them and adjusts probabilities, prioritizations, and cashflow waterfalls. Our advisors then take those goals and process them with risk and other factors to develop an investment portfolio for the couple, much like Google Maps provides various routes to a requested destination.

Traditionally, the concept of excess was previously buried in details and therefore advisors never acted on it. It was a statistical concept that was seen as a probability of success and an overfunded goal. The traditional industry would set meetings based on an annual schedule or centered around a client, and the advisor would only track the goals, with no idea of where the different statistical advantage or disadvantage was—and no advice was being given on this actual statistical dynamic that was occurring.

For example, if the client had $400,000 and wanted to reach $1 million in twenty years with an equity portfolio, the likelihood would be high. If the client because of good sequence of returns was five years in and had reached $700,000, they would have fifteen to achieve the remaining $300,000, so statistically, there would be over an overwhelming chance of reaching that goal. With that certainty in mind, what if you were alerted to the change in probability and was offered suggested places to reallocate it? That amount would be statistical excess because you haven't actually reached your goal yet; you don't have the aspired-to $1 million but you would if you continued to follow the current route. By tracking the client's probability of success, the software points out this excess quotient that can be used for other purposes.

In the past, surpluses were created all the time in clients' financial plans but they weren't being brought to their attention. Instead, clients ended up over-funding goals when, based on statistics, they should've known about and reallocated the excess years before.

Now, rather than wait for a scheduled meeting at the end of a quarter, our advisors contact the client and plan to address the situation with more efficient advice. Identification of surpluses is occurring all the time and our advisors are tracking thousands of goals with the ability to move that excess around more productively. Eventually, as we improve clients' financial literacy, they might do something with that information without even engaging their advisor.

The software also allows our advisors to get what doctors call "a second opinion" on investments. In the past, it had been challenging for financial advisors to do this. While it's common for doctors to have a colleague

> *The software also allows our advisors to get what doctors call "a second opinion" on investments.*

look at a patient's charts and provide a different perspective, the financial services industry has rarely offered this option. There were many reasons and much friction around the issue.

The software is making second opinions accessible. It allows for an advisor to share data with a colleague while still protecting the client's privacy. This can help put together a portfolio that is specially designed for a particular investor. For example, someone might want to explore putting some of their funds in Emerging Markets in Asia, municipal bonds in U.S. cities, or ETFs specializing in space exploration. Their advisor may not have deep knowledge about those opportunities. With the new software, the advisor—or the client—can seek an opinion from another advisor with the needed expertise.

Once again, welcome to the Advice Age! The software enables a client to seek both discrete and/or continuous advice, potentially from different advisors based on their areas of expertise. If one advisor has proficiency in retirement and another has experience in rental homes, both can come together to build a client's portfolio. It is the Age of Collaborative Advice.

Another key aspect of our new software is that the customers own their own data. In the past, if an investor gave information about themselves to a company, the company would own it. In signing up with some of the largest firms, for instance, they would be asked to provide quite a lot of information to that company, and while working with the client, they would collect even more data. And their software would retain that data even if the investor left (compliance, yes). In fact, the client data would be monetized. Our new software, on the other hand, posits the ownership of that data with the client, the individual investor.

There will always be larger firms collecting clients' data and defending the practice by saying it helps the firm give better advice. Does it? The problem is, in the past they started becoming mere merchants of information, data firms—transacting and benefiting off the customers' data itself, regardless of whether the customer was being helped. The ill-logic path that occurred was this: Large firms needed data. Why? To give better advice. So, advisors that are attached to larger firms should give better advice. Do they? Now, in the Age of Advice, these old logics are being reformed. And now, there is a different way, a client-centric way, that is rendering obsolete the many firms that cannot transform.

Of course, government regulations still require the company to keep client information on record for a certain period. Best interest, books and records, etc. Right. But the client ultimately controls their profile of personal details, accounts, investments, and planning statistics. If they're not happy with an advisor at one company, they can take the data and move to another firm. It's the ultimate industry accountability.

The data that our software offers makes the role of the advisor comparable to that of a doctor prescribing medicine. Patients know that there are pills for everything from a minor headache to yellow fever. But it takes a doctor to advise what the best remedy is, when it should be taken and so on, and to provide access to the most powerful treatments. Like exotic out of pocket tests and experimental pharmacology, private investment opportunities are always available for clients. But it's by using our new software, and the knowledge and information it helped gather, that advisors can counsel clients on which investments to make and when to make them. They can also give insights into how the different assets interact one with one another. For example, the use of a specific portfolio to hedge a client's small business exposure.

Today, financial plans require structured solutions like over-the-counter cold medicines that contain multiple products working together, for example, decongestants, expectorants, antihistamines, and others, and there are exciting groupings of products coming into vogue. Based on all the knowledge that clients can provide, companies can offer combination products. This concept was available only to institutions and select clients just five years ago. But even we can do this now. Here are a few that are available to clients: guaranteed income products where premiums are fed from an investment account and paid only when there is excess; inflation-adjusted structured college products that are bought before children are born and can be traded if they don't end up going to college; mortgages that include automatic term life insurance (with an option to reduce payments as the loan obligation declines) which hedges out the full amount of the mortgage so the surviving widowed spouse never has to worry about paying it off—meaning that when you get a mortgage, you can also get embedded new options like insurance or the seeds of a reverse mortgage built in as well.

With the mortgage-insurance combination, what you would be able to draw down would be what you already contributed in. Your equity would increase, which would allow for the income stream to be valued. There is no need for multiple transactions because you're hitting the same data anyway, and all this information resides on the chain. In essence, it's a smart mortgage with multiple feature sets.

The benefit of these combinations is two-fold: Clients get everything they need in a one-stop-shop kind of way, and advisors can make multiple sales. The early opt-out culture allowed regulators to consider things that were previously inconceivable.

But, what if we could take this concept of value one step further? As we've seen lately, the tokenization of nearly everything occurs all the time. In blockchain, tokenization divides the ownership of an asset into digital tokens that have value and are usable on chain. These non-fungible tokens (NFTs) are unique and can represent a work of art, a piece of recorded music, or even your clients. Now, they are their own NFTs because they are unique and have value. Not only are there combination products but, thanks to tokenization,

> *Clients get everything they need in a one-stop-shop kind of way, and advisors can make multiple sales.*

nearly everything and everyone is becoming a financial instrument. Just as we can invest in a music stream from our favorite artist, each one of your clients can be invested in and traded, and that means nearly seven billion new products worldwide. In the DeFi space, individuals are their own combination products and their own cash streams. The fact that money is made from client's data is not bad or evil; however, we believe our clients should permission it and benefit from its use.

As we continue to tweak the new software, machine learning will do the heavy lifting and we will eventually have a planning oversight solution that will allow an advisor to evaluate client portfolio(s), examine their goals, and get alerts on *potential* problems on a twenty-four-hour basis. The launch of ChatGPT fast-forwarded all the possibilities years ago. This new preventative advice with ongoing, substantive solution possibilities and product combinations need a new class of hyper-informed advisor. With the new software, these advisors can also engage customers at scale. In one scenario, one of our advisors might note in *The Wall Street Journal* that inflation may hit high single digits again. They calculate how that level of inflation might affect a client's holdings over two years. They could do a similar calculation with possible moves in interest rates. They would then meet with the investor, present the scenarios, and offer suggestions on how to proceed. The advisor could then repeat that process with ninety-nine other clients. The software would make it easy, tracking the advisor's activity to show the entirety of activity for a given client or book of business, and compliance loves it because it makes it easier for them to oversee the advisor. Businesses have operated with simultaneous contingencies for decades. Now, individual clients can too.

In the new financial advising world, advisors will be portable and scalable. They can pick up and leave if their firm refuses to transform. This puts advisors in the same league as tech engineers. If engineers are done with one firm, they just cross the street and plug into a new one. They don't miss a beat or a paycheck. Advisors are now going to be able to similarly control their destiny.

The traditional dynamic between a financial services firm, advisor, and client will also change dramatically. The center of the relationship between the client and the advisor shifts to the client, and the balance in the link between the firm and the advisor shifts to the advisor. Let's face it; it's hard for large firms to evolve fast. Independent advisors have always been better positioned to follow through with serious planning. As we have discussed, in the Advice Age, the advisor's role becomes even more crucial. Fully equipped with new information around goals, our advisors are better positioned than ever to counsel clients about more innovative investments. They would also know when one set of plans competes with another and what to do about it. And they are poised to craft and adopt structured financial solutions over a period that might stretch from the client's twenties to their eighties and beyond.

As collaborative planning takes hold and more advisors work with this type of software, they could own practically everything about the advising process. The current model of independent advising would change. Eventually, the concept of independence will need to be reimagined.

While this software innovation and the use of the information it provides will undoubtedly transform the financial services industry, the "old school" robo tech will continue to play a limited role. Since machine learning software was first introduced in the early '70s, major firms have used it to their advantage. In fact, machine learning and artificial intelligence are still driving the next wave of the quantitative revolution in our industry. Today, several firms still lure clients with the promise that their investments will be managed by a robo at a meager cost. To their credit, these firms have enhanced the interfaces, added new bells and whistles, and offered new products. They have marketed their technology well. However, in the past, these companies positively drove innovation across market participants. The "winner-takes-all" approach, along with the vertical integration of firms, left advisors with more homogenous offerings, less practical innovation, and fewer controls.

To be fair, algos have helped advance the world of financial services. The innovations succeeded in bringing in younger generations of investors attracted to technology over face-to-face encounters and appealed directly to clients who had extra money to invest but preferred to enter their data into a website and trust the robo technology to run their investments.

In the end, individual investors are the clear winners in the new Advice Age. The new generation of software allows them to receive investment advice that directly addresses their investment goals with a visualization ("planning software") they permission themselves. They have access to prescriptive advisors who are attuned to their changing life situations and poised to manage them. And they have the freedom to move between companies and advisors without the fear of losing momentum.

Think financial goal Rubik's cube.

In this period of modern Copernican innovations, personal advisors and investors can expect further changes. One on the horizon is for the advisor to counsel the investor not only on the right moves to make but also on how to organize the configurations and reconfigurations of product solution sets. Think financial goal Rubik's cube. But enough about software. Let's turn to a different topic and a highly unexpected one; the rather shocking industry shakeout we've witnessed over the last five years.

LEANING INTO THE REVOLUTION: THREE INCENDIARY TRENDS

L et's now turn to those areas in which the revolution has begun but isn't quite completed. You can imagine these trends as three strong initial shocks along industry fault lines, with a new landscape about to emerge. In financial services, there are several such areas worth discussing.

I. Eliminating financial intermediaries

The first trend is *disintermediation*, which is a fancy way of describing the removal of intermediaries when people want to do business with one another. Here's an example: until five years ago, if you wanted a home loan, you went to a home loan lender. If you wanted to put money in the bank, you went to the bank. If you wanted to lend somebody money, you had to call in the lawyers if it was a substantial amount of money; and if it wasn't that much, well, you just took the cash out of your wallet and hoped that one day you would get repaid.

Intermediaries, bankers, and lawyers, etc., serve purposes. They build trust. They protect. They provide guidance. At the same time, though, they add a lot of unnecessary friction, costs, frustration, record-keeping, and delay. If you bought your house five years ago, it probably took you somewhere between thirty and ninety days to get approved for a home loan. Remember that big stack of papers you had to sign at your closing?

Traditionally, life involved paperwork. Remember how annoying it was at the doctor's office? Or filling out the same sheaf of forms (or tapping on iPads, how innovative) every time you went to the dentist? Today, of course, that's ancient history. Thanks to blockchain, if you want a home loan, you can get approved in a day. All you have to do is permission the appropriate information to a potential lender, and you're home free. Well, your home isn't free, but you are.

Similarly, if you wanted to borrow money, you went to the bank and you were subject to that nasty old rule that anybody who needs a loan cannot get one, and anyone who doesn't need a loan can get all the money they want. That's ancient history too, thanks to the peer-to-peer lending phenomenon that also owes its existence to technological advances. So long, intermediaries. You can check out the other person yourself, and if you get back a trust score that works for you, you can get or give a loan. You don't need a bank, and you don't need bankers. *Here is the new rule: Unnecessary intermediaries are obsolete. (There's another side to that rule: Necessary intermediaries are more crucial than ever. But we'll discuss that another time.)* You don't need anybody standing in the way, being a tollbooth, grifting, slowing things down, or failing to return a phone message (anybody remember voicemail?). When it comes to lending money, being frictionless changes everything.

If you drive down Main Street these days, you'll still see "traditional" banks, and you'll still see ads for home-loan originators during NFL broadcasts. That's why I say this is a revolution in progress, as opposed to one that has completely changed how we do business. I would describe this year as part of a transition process, moving from a world of unnecessary intermediaries and friction to a new world, be it lending, borrowing, or what have you; all peer-to-peer or at least friction-free and all brought to you courtesy of - all together now - the blockchain. Just like electric cars, it just takes time for new ecosystems and infrastructure and for people to get used to these things.

If you're old enough, you may remember the 1990s when some people were buying up the names of dot coms for a pittance because they knew the internet was here to stay (nod to Andy Grove and the like), and others thought the internet was an expensive joke that would never have any resonance in people's lives. Those smart people who bought up valuable internet real estate made off like bandits. The rest of us caught on eventually.

> *...moving from a world of unnecessary intermediaries and friction to a new world.*

The same thing happened when utilities laid fiber optic cables in your city or town. You were probably annoyed because workers were chopping up all the sidewalks, and for what, exactly? But man, weren't you happy when you could benefit from fiber-optic technology to enjoy the internet, entertainment, and what have you in the comfort and privacy of your own home? That was also a revolution.

Five years ago, self-driving cars were still in the early adoption phase, like early ETFs in the mutual fund world. Now, they're everywhere. Chances are, if you've got kids in diapers, they'll never learn how to drive! They won't need to, any more than you or I would need to learn to turn the crank on the front of a Model T.

The lesson: things change. All the time. In terms of eliminating the friction, delays, and general frustration that accompany the traditional models of borrowing, lending, investing, planning, and so on, the revolution has begun. Give it a few more years, and we'll pretty much be doing everything on the chain. Just wait and see, and, of course, you read it here first. There will be disruption to the value chain for sure, but there will also be an elevation of financial guidance.

II. Make way for financial coaching

It's interesting that in our conversation around disintermediation, there is another huge change in our industry worth discussing: the rise of a different kind of intermediation: financial coaching. This is a misunderstood subculture. We all know what advisors are, and we all know what financial planners are. But YouTube, TikTok, and the social platforms that allow people to post videos of themselves have transformed the way people learn about investing. There's a tendency for us older folk to look at TikTok and other platforms as places where adolescents post videos of themselves doing crazy dances, hoping to puff up their fragile young egos with gazillions of "likes." But don't discount the power of those platforms to all but undermine the traditional ways folks have received financial advice in this country. Let me explain.

Starting five or six years ago and ramping up with increasing speed, thousands of people post videos (as of this writing, TikTok is only ten years old) explaining different financial concepts to audiences that are, like the Hamilton lyric, young, scrappy, and hungry. These videos, often shot with one camera and featuring "influencers" too young to get a training position at a traditional wirehouse or asset manager, are changing everything. People are putting up free courses, charging money for courses, or simply giving information about how to manage your finances.

The audiences of these financial influencers are young, scrappy, and hungry too. They have practically nothing in terms of AUM to make them attractive to financial advisors. Collectively, they probably have a lot more debt than assets. But these folks are paying $150 to $200 an hour to online financial instructors who are teaching them how to get out of debt, balance a checkbook (not that anyone balances checkbooks these days), or save for a vacation, a car, or if the gods are kind, a down payment on a house. Tons of these influencers, who call themselves coaches, not financial advisors, and, therefore, not regulated, are raking in six-figure incomes selling hourly coaching services to a spectrum of consumers that our industry frankly couldn't care less about.

People are putting up free courses.

When was the last time you saw a bona fide financial advisor lusting after a twenty-five-year-old with a net worth of, say, $2,200, half of which is held in bitcoin?

If you are a twenty-something and you're in credit card debt and don't know how to get out, or you're in your late twenties and particularly industrious and want to own your own home, or you just want to stop being a backward-baseball-cap-wearing knucklehead and get your life moving, then the one to two hundred dollars you spend on a coach is a lot smarter investment than buying six to twelve "lattechinos" at your local Starbucks. Money well spent, indeed. You're learning what you need to know about personal finance, you're getting the information from someone who looks like you as opposed to someone twice your age, and you've committed to a single transactional episode rather than a long-term relationship. If you're a younger person, that's perfect. It's just what you wanted. And now you have it.

It's easy for folks like us to scoff, but the reality is that these coaches are making as much money as many people in our industry who have practiced for two or three decades. This is not just a phenomenon of young people guiding other young people to get their financial lives off on the right foot anymore. Shame on us for overlooking this market, individuals wanting to pay for discrete help or, sometimes, just personalized financial literacy training. Yet, at the same time, if we are honest with ourselves, most of our business models have nothing to offer them.

Coaches are stepping in and filling a gap that we could have filled ourselves but didn't. When these young people have gathered enough money for us to be interested in them, will they be interested in us, or will their coaches have evolved into full-fledged financial managers?

Compliance is finally getting right-footed about a lot of the new technologies and approaches that characterize the world we live in. Because of the FTXs' failure, they're finally getting their arms around cryptocurrency, bitcoin, and a host of other issues that have been the Wild West for the last five years or longer. But coaching has grown, too. Practitioners have no formal compliance requirements. They don't have to run their marketing material by their broker-dealer because they don't have broker-dealers. They seemingly say whatever they want to say, within reason, of course. These are the neo-financial influencers. Therefore you've seen consortiums of coaches get bought up by, or sponsored by, some of the traditional financial services institutions who were smart enough to read the writing on the wall. If you can't beat 'em, own 'em, or at least sponsor 'em. Let them tell their clients about your products and services. It's smart business, and it's happening every day. Further, our firm believes that even if coaching is a fad, discrete financial advice is the trend.

III. The rise of payment processors

This leads us to the third, most explosive of trends I'd like to bring to your attention: the rise of payment processors. Many folks were taken by surprise years ago when that gleaming new football palace in Los Angeles sold its naming rights to SoFi. What the hell is SoFi, a lot of people wondered. To put it simply, they're a payment processor. They handle your student loan payments, your loan payments, and your credit card payments, and consequently, they have a meaningful financial relationship with anybody who has ever taken out a credit card, a student loan, a home loan, or any other form of debt. Doesn't sound sexy, but it makes a whole ton of money.

Over the last few years, whenever I've talked about the rise of payment processors, people have looked at me as though I have a third eye in the middle of my forehead. But, you know, in Hindu culture, that third eye signifies mystical insight and enlightenment. Not too many people accuse me of such vision, especially my spouse. The reality is that the increasing power, reach, and acumen among payment processors has been perhaps the greatest untold story in the financial industry over the last half-decade. I'm talking about companies like Block, Capital One, Chase, Citi, SoFi, Clover, Merchant One, PayPal, Visa, and MasterCard.

Did you know that all the way back in 2021, SoFi had ETF and index products? Or that PayPal explored investments way back during the pandemic and even hired a CEO of Pay-Pal Invest? Why did they do that? Because they've already got a relationship with everybody in the known universe who pays for things. Gazillions of people pay for things on PayPal. So, if you're PayPal, you don't just sit there taking a nice fee every time somebody transmits funds over your platform. Instead, you start thinking about how you can monetize the trust relationships you've created with all those gazillions of nice people out there.

People trust PayPal, which works exactly right every time. When was the last time you didn't get your money via PayPal? When was the last time you sent money via PayPal, and it went where it wasn't supposed to go? That's trust. Monetizing trust for PayPal meant monetizing a suite of investment opportunities to its myriad customers. And tons of those customers took them up on that offer. As a result, PayPal, SoFi, Block and all those other entities who seem to be doing nothing more than boring, back-office transactions, picking off a couple of points on fees in a decidedly unsexy way, suddenly became the belles of the ball. If you're a financial services firm, there's an awfully good chance that a payment processor is going to try and buy you, be you an advisor, broker/dealer, bank, investment firm, etc. I am still watching for Visa and UBS to get together or SoFi to buy Wells Fargo Advisors. We have already seen a few transactions. They're also going to buy one of those consortiums of coaches I mentioned

> *I am still watching for Visa and UBS to get together or SoFi to buy Wells Fargo Advisors.*

earlier. Thus, payment processors will have it both ways; they will offer advice in the traditional sense, through stodgy old folk like us, and they'll offer wisdom and guidance on an hourly basis through the army of coaches they now own or subsidize.

When we think about these initial shocks to industry fault lines, we think cinematically: Great cracks in the earth, elevated highways crashing to the ground, fires everywhere. I'm not talking about those kinds of scenes. I am talking about massive changes in our industry in these three areas. The increasing disappearance of unnecessary intermediaries and friction, the rise of the class of online influencers/coaches and discrete advice, and the rise (and rise!) of payment processors. These shocks appear to do nothing more than rattle your dishware, but in truth, they are signs of a realigning planet. A lot of people are missing out on these three trends, but not you. Now you know all about it.

CONSOLIDATION HITS HOME

L et's turn now to another of the burning issues in our industry that affects both advisors and clients alike.
I'm talking about consolidation.

To put it simply, many of the big-name firms whose presence seemed eternal in our industry have now disappeared. Larger firms have gobbled up smaller ones. The result has been enormous dislocation, not just for advisors but also for clients who have found themselves and their assets transferred from one firm to another, often multiple times. Why did this happen, what are we to make of it, and what can we expect going forward?

Herb Simon, an economist who worked in the '70s Nixon White House, uttered these brilliant words: "Things that can't go on, stop." In some sense, it was inevitable that the changes we've seen in our industry would take place. However, most advisors, and many firms, were blindsided because they thought they were immune to the sorts of economic pressures that govern pretty much every other industry and economic sector in society.

Let's start by talking about roll-ups. As we all know, a roll-up is a firm that buys multiple companies in a given space to create a dominant market share and increased economic power. Sometimes, this happens to the benefit of consumers; sometimes, it's to their detriment. But we've seen the roll-up phenomenon in countless industries. Many of us in financial services have played a prominent role in funding and otherwise abetting them.

Everything from HVAC firms to companies that make window shades (an incredibly lucrative business; I just saw the bill for window shades in our new home) and from local TV stations to entire cable networks have been subject to roll-ups over the past twenty years. The remarkable thing is that whenever the roll-up phenomenon touches an industry, pretty much all the participants say the same thing: "I've heard of these things, but I never expected that they would happen in *our* industry."

There's a tendency for people to believe that if things are a certain way, they'll always be that way. And yet, we financial advisors with front-row seats to economic change and what Joseph Schumpeter called "creative destruction," we who are often responsible for such massive changes in countless industries, never saw it coming when the roll-up phenomenon reached our office doors.

And yet it did, as per the Golden Rule: Whoever has the gold, rules.

In our case, just a few short years ago, thousands of smaller, not quite mom-and-pop but decent-sized advisory firms received *The Godfather*-like offers they could not refuse. They were able to merge their assets, teams, and technology with larger firms for ungodly amounts of money. The offers were simply too enticing and created opportunities for firm owners to take a lot of chips off the table while still maintaining an ownership interest and some control.

> *...advisory firms received* The Godfather-*like offers they could not refuse.*

But all too often, some firms became so successful that no one, including the owner's adult children, could afford to buy them out. The owners became victims of the firms' success. As a result, we've seen thousands of such firms disappear, combined into a handful of new mega-firms whose names none of us would have known just a decade ago. The aggregation phenomenon continues unabated to this day. There scarcely remains a small firm with, say, under $250 million of AUM whose owners have not been solicited numerous times by the roller-uppers. You've got to be extremely stubborn, extremely rich, or extremely foolish not to take advantage of the seven- and often eight-figure checks these mega-firms are writing to the partners of these smaller firms.

But it wasn't just the money that made the roll-ups' offers so attractive. Firm owners read the writing on the wall and realized it would become increasingly difficult, if not outright impossible, to compete with the technology, range of services, lower fees, and broader selection of investment opportunities the new mega-firms offered. As they say in the military, "Sometimes surrender means joining the winning side."

This is one of the two main areas in which the landscape has changed mightily in our industry: the consolidation of smaller firms into larger ones.

The other equally unimaginable trend over the last five years, which no doubt will be hastened by the development of the technology I described in the previous section of this letter, is the disappearance of certain big brand names. Every football fan knows them, having seen thousands of ads during breaks in NFL games, where these firms used to advertise heavily.

I won't be too generous with names because I don't want to seem like I'm dancing on the graves of our competitors, but I find it as shocking as you that so many of the big-name brands, the ones that we considered "too big to fail," no longer exist. RIP E-trade, TD, and Wells Fargo Advisors. How could so many of the tallest trees in the forest have come crashing down?

In a transitioning industry, the lack of innovation is always a concern when it comes to the life and premature death of an enterprise, no matter how big or small. And it's not hard to look back to see that much of what was happening in dearly departed firms really smacked of problems with decision-making. For example, one firm thought it would be a great idea not to build a data warehouse to extract the demographic trends of the underlying books of business of their firm. They did not institutionalize client retention strategies, risk management, or modern digital marketing to gate and grow customers. And so, one of their competitors targeted their top advisors' clients with a sophisticated digital campaign and won. It is a jungle—you eat what you kill. Regardless of whether we acknowledge or engage in it, a constant war wages for the end customer (our customer). In other words, advisors expect their firm to defend—in depth—the success they create for themselves, their clients, and, lastly, their firms. I worry that some advisors may not even know what is protecting their life's work, and others overestimate it.

At the traditional, larger firms, it really didn't matter as much how considerable the value you created for your clients. All that mattered to Corporate was how your number benefited the firm's bottom line. That trend certainly fell into the category of things that went on until they stopped. It's pretty much impossible, as the last five years have demonstrated, to keep advisors working hard for a firm that fails to match the compensation of its competitors and centers all its energy around the customer.

As a result, the better advisors departed firms with the employee model, and they departed in droves. They left behind the advisors who really weren't good enough to cut it elsewhere, who lacked any sort of entrepreneurial spirit, who didn't believe in themselves, or who were just marginally competent. Running a firm where a preponderance of your advisors falls into that dismal category is obviously a recipe for failure. It's like going to a restaurant because of its past reputation only to find that what had distinguished it—the head chef and most of its staff—was gone.

And then, of course, the planning software that we discussed in the previous section had a lightning effect on our industry. We've talked in this letter about the two Copernican revolutions—one that puts the client at the center of things vis-à-vis the advisor, which is a function of new industry plumbing: blockchain, and one that places the advisor at the center of the universe vis-à-vis the firm, which is a function of the planning software.

When your advisors can up and leave at any moment because you lack the ability to put them on "garden leave" or otherwise lack the ability to handcuff them and their clients to their desks, you create an environment where, with Darwinian cunning, the strongest survive. Who are the strongest firms today? Who gives the best financial advice today? Not the ones that exert the most power and influence over their advisors. As we've seen, those have become the dinosaurs that looked up to see a meteorite coming and never had time to look back down. Instead, the firms that have survived the shakeout so far, and the ones that will be here five or ten years hence, are the ones who make it easiest for their advisors to serve their clients effectively. You will be happy to know it is firms like ours.

Who are the strongest firms today?

In some ways, this is a reprise of what we saw several years ago when the regional firms began to snap up large numbers of advisors and massive amounts of AUM from their bigger and better-known competitors. Some advisors were so entrepreneurial that they relished the challenge of putting together their own firms from scratch. That meant hiring the IT guy and fixing, when necessary, the copier, doing all the compliance work, and so on. Many such advisors created their own firms and thrived, and large numbers of them are the ones who participated in the roll-ups I described earlier.

Yet most advisors who chafed under the restrictions of the larger firms felt some degree of entrepreneurial spirit but did not want to be the person who called for the copier to be repaired. For them, the regionals, independents, and roll-ups provided a safe landing place. Those firms offered all the technology, all the support, all the communications tools, all the compliance teams and all the products advisors needed without the red tape, bureaucracy, or hierarchy, but with the mantra of, "if we make an exception for you, we'll have to make an exception for other people, too."

Many advisors are innately gunslingers. They don't like being told what to do or how to do it. Firms like the independents provided safety and security along with the entrepreneurial freedom and ability to run one's team as one saw fit, which appealed to an increasing number of top advisors. Firms like Cambridge are doing some wonderful things. This phenomenon hastened the demise of some of the old guard whose names were etched into the minds of football fans everywhere because of the millions of dollars spent on ads—now all for naught. On the other hand, some of those regionals can no longer be called "regionals" because, in some cases, they grew bigger than the national firms they displaced. They grew so big, fat, and happy that they found themselves subject to some of the same fatal problems that took down their better-known competitors.

And then you have the robo advisors, which turned out to be far less of a threat than we assumed ten or thirteen years ago when pretty much every advisor believed that robos would eat their lunch, and their breakfast and dinner. We were thinking trillions in AUM. Instead, robos learned the hard way that they could not provide financial advice effectively and, in a twist of fate, had to adopt the human approach. Consequently, the clients with the really big money, the ones whose assets were meaningful enough to attract the top advisors, might have dallied with a robo, might have even opened an account or two, but ultimately came rushing back to the safety and security of a human advisor. Perhaps their new partnerships and configuration with AI will give them life once more.

Of course, robos were just traditional players who had come into a traditional market with a cool, digital front end—a slick front end positioned to solve the onboarding problem that we previously discussed. Their differentiators were marketing and price. Many analysts exaggerated how powerful robos would be and ignored the real threat right around the corner. Surprise! It's payment processors that have changed the industry forever.

The ascent of payment processors reminds me of a well-known Harvard Business School case study involving Steinway and Yamaha pianos. Many people knew the Yamaha brand for its motorcycles. At some point, Yamaha decided to start making pianos. Steinway, one of the world's premier piano manufacturers, wondered if Yamaha's pianos could be a threat to its business model. Many of the company's analysts and consultants told the brand not to be concerned. Wrong! Soon, Yamaha reverse-engineered Steinway's pianos. By using similar quality materials and turning to automation to bring down the price, they attracted younger customers with smaller wallets and nearly put Steinway out of business.

How does this relate? The real risk was not that robos were a potential replacement for advisors, it was that payment processors, using a different technological backbone from another

> *Soon...robos began to beat traditional asset managers and advisors.*

part of the industry, could quickly build fully integrated offerings and attract customers. Soon, they began to beat traditional asset managers and advisors at their own game. While advisors only dealt with the asset side of the balance sheet, payment processors came in via the debt route with younger players than those the larger firms relied on.

At that time, financial advisors weren't interested in trying to make money off younger people because they didn't have any money to invest. Instead, they targeted more mature investors at the top end of the market.

Now, payment processors are transforming the industry because they are sitting on so much new technology that they can siphon off and monetize younger clients in droves. As these clients age and gain wealth, they are already locked in and loyal.

When payment processors first came on the scene, they were a cross between software engineers and digital marketers. However, as they've become successful, they have started to add advice and advisors, and some of the most interesting products are now not built by typical asset managers but by payment processors. Their cross-sell opportunities and retention are higher than any traditional player can boast because they have investment, banking, lending products, and planning, and they have built out their tech infrastructure. That means they can do things a lot faster and at a lower price. They can offer more product channels faster than traditional players can. They even offer AI-driven product bundles. It's truly amazing how much payment processors are starting to take the lead in asset management because they've figured out how to grab clients earlier in the financial advice lifecycle.

The opt-out community has rushed toward this new market participant. The opt-out community and payment processors are forming the new industry model, and many large firms are now faced with an "adapt or die" strategy. As of this writing, we are trying to figure out how to compete in this new reality. Robos were never the real issue, and we focused on them for too long when the real competition was right in front of us. In fact, I have checked my wallet, and I personally use several of our new competitors. Let's just say that no one thought Yamaha could be a threat to a firm like Steinway until it was. Payment processors have the Steinway piano laid out in their warehouse, they have reverse engineered it, and they are building their own version with all- new technology.

One of my themes throughout this letter is that you cannot provide advice at scale, no matter how sophisticated your software may be. Robos thought they would win the war without firing a shot, and many top financial services believed they would lose the war. And yet, as we've seen, the payment processors of this world have done a tremendous job serving the broad swath of folks in our society whose assets are limited, who suffer from credit card debt, who are still trying to buy their own house, and who want to be one of the ships riding at tide instead of mired in the mud at the harbor entrance.

So, hats off to all payment processors. But when it comes to managing and advising, and I use the word "advising" advisedly, affluent individuals and families with meaningful net worths, processors are still enhancing their solutions. We didn't know it then, but we know it now. However, we were, and are, enhancing our solutions, too.

Where do we go from here? It's entirely possible, given software advances and a more intelligent compliance structure, for advisors to band together and create their own neo-advice type firms. But the financial services universe today is dominated by the biggest players, the firms that adapted and survived the transformations and shakeouts of the last few years, and the firms that intelligently and effectively rolled up what I'm affectionately, not derisively, calling the moms-and-pops.

Is there still room in the world for smaller independent firms? Yes, absolutely, if they're nimble and treat their advisors with respect, recognizing that advisors today can find greener pastures and take their clients with them at the drop of a new edict or policy from the corner office. I expect to see this trend accelerate with a smaller number of big firms remaining, with a few more tall trees still to topple, and with independent firms succeeding as never before. Because today, thanks to blockchain, you don't even need a copying machine.

No one ever expected roll-ups in financial services, but no one ever expects roll-ups in any industry. They just happen. And no one expected those tall trees to come crashing down, but they did. The future belongs to the firms that treat their advisors right, that recognize that advisors, like software engineers, can leave one firm today and start somewhere else tomorrow, and there's not a darned thing the prior firm can do about it. The power has shifted from the firm to the client and from the firm to the advisor. It's liberation for advisors who can run their professional lives as they see fit to a greater degree than at any time in the history of our industry. We welcome that transformation, of course. We've helped hasten it by being part of launch planning software that created portability for advisors. And we work hard to make our firm a place where you, our advisors, are happy and free to do their jobs and personalization is not a gimmick to siphon more data to sell or build product moats around our business.

To the victors belong the spoils. So, here's to a brave new world of creative disruption in financial services, which, at its core, means that financial advisors are freer than ever before to do what they were supposed to do.

Advise.

Wealth Management Industry

Past

| **Payment Processors** *Debt Market* | **Traditional Wealth Management** *Asset Market* |

Present

| **Payment Processors** *Expanded Offerings* | **Traditional Wealth Management** *Losing Market Share* |

Near Future

| **Payment Processors** *Expanded Offerings* | **Traditional Wealth Management** *Pushing Back* |

Long Term

The Future
Consolidated Debt and Wealth Management Market

THE DEATH OF DRAG: ADVICE, PERIOD

W̲e achieved one final, notable accomplishment in the past year: getting rid of drag. What's drag? We're not talking about highly costumed gender-benders. If you're a client and the term is not part of your everyday vocabulary, you've undoubtedly experienced it anyway. Drag is the set of slow processes and tasks that advisors have had to do while working with clients. It's friction. It's an umbrella term covering interactions with the home office, interpreting and enforcing regulations, filling out paperwork, adapting to technology, and the many other actions that advisors have had to perform behind the scenes.

Well, at our firm, we've eliminated everything in the drag category. The move took us another huge step forward in creating a seamless way for our advisors to work directly with investors. It has been a long, hard, continuous digital transformation, as deep learning is adding new, undiscovered places to add value, even now. And there are many advisors still trying to figure out what the "right" tech stack is for them to use. But that is the wrong conversation. It's actually about moving from product sales to planning to personalization. That personalization must include putting our clients at the center, not only with regard to advice but also to the controls over the data that their advice is based on. Further, control means teaching our services in a way that our clients are empowered to trust but also verify our advice.

Doing away with the drag may be the last, but certainly not the least, of the changes this modern Copernican era has brought to the way we operate as financial advisors. It represents just as significant a sea change as introducing passporting and discrete advice or the other new approaches to advising that we've rolled out.

Drag has long hindered the efficiency of the advising process. The pressure drag put on the way advisors operated went beyond cumbersome levels of advisor/investor communications that we

> *...clients are empowered to trust but also verify our advice.*

discussed elsewhere. Yes, we also did away with those unnecessary bureaucratic stages of advisor-client engagement.

You could compare it to what happened with computers before the turn of the millennium. For years prior, everyone went through life happily using their HP desktop or RadioShack laptops (ask your grandparents what the TRS-80 was), snagging tech upgrades and so on. Then, Y2K came along, alerting the world to all the glitches that were built into computers. That was the computer system's drag moment. People worked so hard to eliminate the glitches that the rest of the world thought the idea of a Y2K debacle was just a myth.

Similarly, a couple of years back, the financial services industry became aware that we'd been saddled with our kind of drag for a long time. And that drag, in turn, was taking away the time and energy that advisors had to work directly with clients and fulfill their investment goals.

For years, the step-by-step moves toward using more sophisticated tech at most financial services firms made the drag load even heavier. Many tech innovations were designed to do some of the tasks that advisors had been doing. For instance, instead of sitting down with an advisor to discuss their risk tolerance, a client only needed to log on to a computerized form and tap in answers to a few questions. Seconds later, they would receive an analysis of the level of investment risk they could bear.

Perfect, right? Well, maybe not so much. How about when the investor felt the computer-generated risk level was too high or too low? The firm still needed advisors as much, if not more, than ever, because an advisor would have to enter the conversation and make the necessary adjustments. Vanguard was an early adapter to tech but still ended up hiring advisors. They used and adopted robo at scale but built out their advisor corps as well. And let's face it, if anyone could have eliminated advisors, it would have been Vanguard, and yet, they didn't and couldn't. In fact, advisors became a fixture of their business model.

With this in mind, don't be surprised when payment processors start buying large broker-dealers to infuse advisors and advice into their business models. Any day now, PayPal could buy, say, LPL Financial, or Capital One could merge with Raymond James. It's just inevitable.

With tech innovations coming and advisors juggling to adopt them, along with their other duties, tensions emerged. That, too, added to the drag. A decade ago, financial advisors boasted a heavy concentration of males in their forties and fifties. Some were early adapters to the tech tools that were becoming commonplace. But many struggled with integrating tech into their approach. The more firms brought on technological innovations, the more some were challenged to figure out ways to pivot. Even after adapting to the changes, they still had to find ways to articulate to clients how it all fit into their investment plans. This kind of tension was a big part of the drag.

The transformative changes that we discussed in other chapters paved the way for us to do away with the drag. First, our blockchain adoption allowed us to collect all an investor's critical data in one place and make it easily accessible to advisors. Then, we rolled out our new software. Among many other advantages, our planning innovations allowed us to evaluate goals in real time. These changes made the dialogue between advisors and clients easy and seamless. Many of the old ways of advising were what had created the drag in the first place. As a result, once we made these changes many aspects of drag were automatically eliminated. Next, we reviewed our advising system, pinpointing where the remaining drag was, and got rid of that, too.

One of the first big changes that resulted was doing away with the intermediaries in any given transaction. In the past, when a client made an investment decision, the accounting on the original transaction would pass through several different hands— middlemen and what we call "rent-seekers." The process was analogous to the sausage-making that goes on when politicians want to create new legislation. We had similar procedures in financial services. A custodian here. A service manager there. Each would have their fees, all passed on to the client. Worst of all, the process was not transparent. Thankfully we, and much of the rest of the industry, have cut way back on all that. Our process is much cleaner, more transparent, and middleman-free.

Let's take a close-up look at another way wiping out drag has lightened advisors' load: cutting back on the paperwork they have to wrestle with. Shifting from asset-focused to goals-based advising is one of the things that has made this possible. When advisors focused on investments, they were required to collect reams of disclosure forms and explain them to the clients so they could feel comfortable enough to sign them. The reams of disclosure were mainly needed because there is so much jargon in our industry. And moving to goal-based investing calls for more descriptive terminology. Clients and advisors interact in a goals-based framework, where the concept is understood because the language is image-based. The whole world has pivoted to images, from Amazon to Instagram. With investment-based advice, we were stuck describing graduate-level mathematics. So, we had to pivot as well.

Of course, goals-based advising requires disclosure, too, but far less than the investment-based approach. This requires the advisor to fill out far fewer forms and allows a lot more time to focus on creative ways investors can reach their goals. Funny enough, the product bundles are actually more sophisticated.

Saying *adios* to drag also helped ease the regulatory pressure on advisors. This had been building up for years. Let's say an advisor was trying to set up a 401k, an IRA, or even something more straightforward like establishing a taxable or nontaxable account. They had to be an expert, not only at giving financial advice but also in how the regulations applied to every specific client as well as each of these and many other vehicles they were offering. Without the drag, these interactions are now far less complex. Taking away the drag—and

> *Taking away the drag...has given our advisors far more time.*

eliminating the inefficiencies that came along with it—has given our advisors far more time. Now, they are going to have to figure out how to spend it. They can go in different directions. Our approach, like many, has been to do more and better planning, specialize and spend more time in meetings with clients. This would be ideal for those advisors who prefer a hands-on engagement approach. Or, now they can take more time off than just Fridays—but when they come back they will find the assets and clients gone and a "thank you" note from all their competitors signed: Obsolete.

Getting rid of drag has been just as positive for investors as for advisors. Our investors are relieved to be done with the gobbledygook lingo that was part of drag. In many traditional firms, advisors typically used "Advisor Speak," the financial equivalent of Shakespearean English. Okay, they were not quite as bad as saying, *Wouldst thou consider a possible ETF consolidation?* But words like "optimization," "beta," "standard deviation" and "risk premium" were rolling off their tongues. As firms transitioned from traditional advisors to a goals-based system, advisors were still talking as if they were on the border between one country that spoke Chaucerian English and another that used a Brooklyn dialect. They'd use bits and pieces of the tongue of each world, then, they'd slide into corporate terms like "cash flow" and "balance sheet."

Now that we've shed all the drag and fully adopted goals-based advising, you find advisors communicating with clients in more straightforward, simple, modern terms. "Hey, we need to save more," they might say, or "You will run out here," or "Your excess is this," or "The tradeoff is that." Naturally, the investors we work with are thrilled with this change. In particular, the younger generation of investors, who were never comfortable with the "old school" lingo of investment anyway, are delighted.

Probably even more than establishing a better common language, investors are pleased with something else that eliminating the drag has brought: more clarity around the value that advisors bring. All of our advisors are peer-ranked and remain in the top decile. Congratulations!

Firms have never had an easy time enumerating all an advisors' tasks and articulating the added value to investors. Should advisors' role mostly be cast in terms of the face-to-face time they spend with clients? Or should it include the time spent analyzing a client's portfolio and coming up with recommendations? What about the time spent hiring and training their client service staff? The drag only made capturing what an advisor actually does even more complicated. How do you summarize the costs of learning new technology, filling out paperwork or meetings with executives at the home office? And once you've calculated all the advisor's tasks, should they be compensated through commissions, hourly or in some other way?

A decade ago, Vanguard published an article about the value of advisors, positing that advisors are worth only a fraction of the percentage of assets they charge. Vanguard was basically saying that planning was worth thirty basis points; that is what they charge for advice and that's the true value of advisors. (Typically, advisor's fees are about one percent of the investor's portfolio per year.) The article said that investors who work with advisors benefit from a three percent additional annual gain, on average. The study then used the one percent cost versus the three percent gain as a marketing point for investors to work with advisors. A couple of other big firms came out with similar studies.

However, this positioning did not go down well with advisors. And no wonder. It brushed over all the advisors' tasks, including all those the drag required them to do. It relegated the advisor to the role of a mere marketing tool, not an expert or a professional but a *salesperson* or *relationship guy* who's only there to emphasize with or offer sympathy to the client. As we all know, there's much more complexity and planning involved than these terms suggest. What's more, the articles and studies were based on views that advisors were ensnared in a web of old technology. Now, new technology has freed advisors of these entanglements of the past. Further, who gives Vanguard's clients a second opinion?

In the Age of Advice, technology propels advice to become the centerpiece of the client relationship. When Vanguard originally published the article, the industry wasn't set up for advice, but it is now. We no longer need to worry about our time with document prep and storage or with constant compliance issues. Instead, we can focus on our advice, and our advice is worth substantially more than what Vanguard said back then. In fact, clients search out advisors because of what we do: give advice.

Getting rid of the drag has brought clarity to what advisors do. They advise. It's even simpler for our advisors. They help investors pinpoint their investment goals and find ways to meet them.

Firms have different views on the best way to compensate advisors for their work. The subscription model, in which investors pay a flat rate for a range of services, is gaining popularity. The subscription model (AUM) could work best in scenarios where advising occurs daily. However, many firms prefer subscription and hourly models. What's not in doubt is that investors who work closely with our advisors will get their money's worth.

The elimination of drag has moved us farther from focusing on portfolios and planning and closer to personalization in financial services. Drag created barriers to the personalized relationship that many investors want. For one thing, it made transparency more difficult. For another, communication was more complex. And, as I have shown, drag also consumed hours of an advisor's time. With all those barriers removed advisors can focus more on individualized goals-based planning. Now that the advisor/investor relationship is more personalized, it can be finely tailored to each investor's goals. With the advisor now

> *Planning can be...*
> *pinpointed on a single*
> *investment purpose.*

engaged in conversations about planning, these communications can be refined and precise to different levels of planning, ranging from a comprehensive approach to one that is pinpointed on a single investment purpose.

LESS COMPLIANCE AND, PARADOXICALLY, LESS THEFT

E very so often, I'll see a news story about how some low-life advisor managed to steal $100 million from his clients to fund an outrageous lifestyle, often involving Lear jets, inappropriate relationships, and illegal substances. I always ask myself the same question, "How did they get away with it?"

I'm not asking because I have a secret desire to enter a life of criminality funded by our clients' AUM. Instead, I'm wondering how it's possible today that the (very few) bad actors in our industry can still steal so much money from their clients. It doesn't make sense. Don't we have compliance procedures that have been carefully honed since the discovery of Bernie Madoff or Sam Bankman-Fried? Don't we have the SEC, FINRA, and even the Department of Labor weighing in and adding to advisors' compliance burden?

We do. And yet, bad people still do bad things. Sometimes, they get away with it; sometimes, they don't.

When I was a kid, our next-door neighbor was burglarized. This happened at a time and place in American history where home invasions were completely unheard of. The family ended up installing the first alarm system, complete with bars on some of the windows, that any of us had ever seen. My buddy lived in that house, and I remember his mom saying, "It feels like the bad guys got away with it, but *we* went to jail."

That's how many honest advisors, which is to say, almost all advisors, feel about compliance today. It's a nuisance, it's a hindrance, it's a complete waste of time, it keeps them from spending time advising, and it does little to keep the bad guys from stealing that proverbial $100 million. So, the question naturally arises in a letter like this, "What's the future of compliance?"

The simple answer is that compliance is going to get a lot easier for advisors to deal with. With blockchain, we're not going to have written signatures on documents pretty much ever again. That means it will be harder to forge signatures, which is a key to how you steal $100 million (not that you asked). Since all relevant documents will be on blockchain, permissioned by their owners only when necessary, the bad guys simply won't have access to the documents they need to steal funds. In other words, security is built right into the transaction, minimizing the need for the kind of complex compliance regime that we have today.

It's a little like robbing a bank. Back in the day, banks had huge amounts of cash on hand. A bank robber could really get away with a lot of money. Today, banks keep considerably less

> *What's the future of compliance?*

cash in their vaults, partly for security reasons; partly because much of our economy runs on credit cards, ACH, and other non-cash payments; and partly because, thanks to AI, they know how much cash they'll need on hand pretty much every day. So, robbing banks is both difficult and at the same time, far less lucrative.

Not that I recommend bank robbery, either.

What I am suggesting is that, going forward, financial institutions will be more like banks that don't have a lot of cash lying around. If it's harder to access money to steal, then there will be fewer bad guys stealing money. What's that line about robbing banks? "I rob them because that's where the money is." That was bank robber Willie Sutton's explanation for his career choice. Well, Willie and his ilk won't have much to do in financial services, with blockchain and other protections, so your compliance burden will be far less going forward.

And ain't that even more good news?

FINAL THOUGHTS: BUCKLE UP

I told you at the beginning of this letter that the past year has been all about revolutionary change. And was it ever! We as a firm took time to define what financial advising was and clearly what it was not. We distinguished financial advice from investment advice. If you have read my letters in the past, I took time discussing what an industry transition would look like, describing it as an iPod moment. We then took years to talk about change management and adopting disruptive technologies. We even voted to move from a lifestyle business to a business that will have a more sustainable path because we agreed that lifestyle businesses have no incentive to change.

And so change is here and we are ready. So we have expanded our use of blockchain. We've created a whole new software focused on the goals of investors, product bundles and factorial solution sets. Think Google Maps. Personalization is creating these "routing" solution sets, so while there are not infinite solutions, there is a very large or "factorial" set of them. We have paved the way for direct engagement between advisors and investors. And finally, we have brought about a drag-free advising climate, including a meaningful reduction in the compliance burden.

Yes, the Advice Age has arrived.

We can't wait to see what next year brings. So, here's my advice: Buckle up! The changes are not going to stop. The one thing I can promise you is that the wild ride continues, but it will be better than ever for advisors and clients alike. A happy and healthy 2028 to you, your family, and your bottom line.

ABOUT THE AUTHOR

Robbie Cannon is a fintech investor in firms such as New Retirement, OnTrajectory and Kinnect. He is the founder of Horizon Investments. He currently serves on the board of Horizon Investments and Founders Financial. He is active in Money Management Institute, a New York based industry association.

During Cannon's tenure as CEO of Horizon, the company grew from a small retail registered investment advisor to a multibillion-dollar third-party investment management group with a national footprint of independent broker-dealers and institutional clients. His career in financial services has encompassed various aspects of investment management, product development and risk management. Under Cannon's leadership, Horizon was nationally recognized. Some of these awards include: Inc 500/5000 for three year revenue growth in 2020, MMI for Innovation in 2014, Asset Manager of the Year in 2018, 2019 and 2021 ($25 Billion & 10 Billion in AUM or less), Top 10 Wealth Management Technology by Banking CIO Outlook in 2021 and Asset Manager of the Year and Strategist of the Year by Envestnet in 2018

Cannon is a national speaker on topics including future industry trends, goals-based investment management and market dynamics. He graduated from Furman University. Cannon and his wife live in Charlotte, North Carolina and have three sons.

www.ingramcontent.com/pod-product-compliance
Lightning Source LLC
Chambersburg PA
CBHW041916190326
41458CB00048B/6839/J

.